SpringerBriefs in Applied Sciences and Technology

SpringerBriefs present concise summaries of cutting-edge research and practical applications across a wide spectrum of fields. Featuring compact volumes of 50–125 pages, the series covers a range of content from professional to academic.

Typical publications can be:

- A timely report of state-of-the art methods
- An introduction to or a manual for the application of mathematical or computer techniques
- A bridge between new research results, as published in journal articles
- A snapshot of a hot or emerging topic
- An in-depth case study
- A presentation of core concepts that students must understand in order to make independent contributions

SpringerBriefs are characterized by fast, global electronic dissemination, standard publishing contracts, standardized manuscript preparation and formatting guidelines, and expedited production schedules.

On the one hand, **SpringerBriefs in Applied Sciences and Technology** are devoted to the publication of fundamentals and applications within the different classical engineering disciplines as well as in interdisciplinary fields that recently emerged between these areas. On the other hand, as the boundary separating fundamental research and applied technology is more and more dissolving, this series is particularly open to trans-disciplinary topics between fundamental science and engineering.

Indexed by EI-Compendex, SCOPUS and Springerlink.

More information about this series at http://www.springer.com/series/8884

Mohd Azraai Mohd Razman ·
Anwar P. P. Abdul Majeed ·
Rabiu Muazu Musa · Zahari Taha ·
Gian-Antonio Susto · Yukinori Mukai

Machine Learning in Aquaculture

Hunger Classification of *Lates calcarifer*

 Springer

Mohd Azraai Mohd Razman
Faculty of Manufacturing and Mechatronics
Engineering Technology
Universiti Malaysia Pahang
Pekan, Pahang Darul Makmur, Malaysia

Rabiu Muazu Musa
Centre for Fundamental and Continuing
Education, Department of Credited
Co-curriculum
Universiti Malaysia Terengganu
Terengganu, Malaysia

Gian-Antonio Susto
Department of Information Engineering
University of Padua
Padua, Italy

Anwar P. P. Abdul Majeed
Faculty of Manufacturing and Mechatronics
Engineering Technology
Universiti Malaysia Pahang
Pekan, Pahang Darul Makmur, Malaysia

Zahari Taha
Faculty of Manufacturing and Mechatronics
Engineering Technology
Universiti Malaysia Pahang
Pekan, Pahang Darul Makmur, Malaysia

Yukinori Mukai
Department of Marine Science
International Islamic University Malaysia
Kuantan, Malaysia

ISSN 2191-530X   ISSN 2191-5318   (electronic)
SpringerBriefs in Applied Sciences and Technology
ISBN 978-981-15-2236-9   ISBN 978-981-15-2237-6   (eBook)
https://doi.org/10.1007/978-981-15-2237-6

This Springer imprint is published by the registered company Springer Nature Singapore Pte Ltd.
The registered company address is: 152 Beach Road, #21-01/04 Gateway East, Singapore 189721, Singapore

Contents

Chapter 1
Introduction

Abstract This chapter starts by exploring the motivation behind identifying fish hunger behaviour. The elaboration on factor triggers fish behaviour which will be explained specifically towards hunger characteristics. The implementation of technologies using image processing to extract significant parameters will be discussed. Lastly, the machine learning (ML) techniques are used in fish behaviour for classification. The outcome of this chapter is to recognize the underlining framework by combining aquaculture, engineering and artificial intelligence (AI).

Keywords Fish hunger behaviour · *Lates calcarifer* · Machine learning · Classification · Image processing · Aquaculture

1.1 Overview

The sustainability of sustenance is very crucial more importantly, with the ever-growing population that increases food demand yearly. By managing the food supply or protein source specifically, the amount of reared fish must be kept concurrent with the resources consumed. Table 1.1 demonstrates the projection of fish captured and reared aquaculture predicted by the Malaysian National Food Agency Policy 2015–2020 [1]. The overall trend suggests that the total amount rises gradually with the apparent escalation from 711.06 to 1,443.00 metric ton for aquaculture from 2016 to the projected year of 2020. The amount shows the upsurge to nearly twice the amount from 2016, and this implies that the requirement of providing enough supply of fish for the incoming years is inevitable.

One of the solutions in providing fish supplies is by increasing the number of reared fish by fish farmers. However, the relation between the number of reared fish and efficient feeding process is highly correlated [2]. On the pretext of feeding, there are several drawbacks by using the normal method in dispersing food which is the time-based feeding and the amount according to farmer experience that would lead to water contamination or food surplus should there be any overfeeding occurred [3]. The water quality reduces as the wastage food dissolves due time, and this would result in the decreases of oxygen that could cause diseases in fish [4, 5].

Table 1.1 Fish production in Malaysia

Year	2014	2015	2016	2020
Capture	1,458.10	1,415.00	1,455.00	1,577.00
Deep sea	1,193.00	1,100.00	1,100.00	1,100.00
Coastal sea	260.50	315.00	355.00	477.00
Aquaculture	520.51	628.21	711.06	1,443.00
Freshwater	106.73	169.75	189.85	313.00
Fish	106.33	169.00	189.00	311.65
Brackish water	168.45	168.41	191.21	400.00
Fish sea	245.33	47.54	61.04	165.86
Total	4,058.95	4,012.91	4,252.16	5,787.51

Most of the existing demand feeder systems lack of feedback input that characterize the behaviour of the fish that is being observed. This is because it only disperses the food according to the pre-set time or by triggering a sensor, and the volume was predetermined by the capacity of the feeder mechanism. [6–8]. However, the projection of understanding the relation between fish behaviour and feeding routine was not integrated as this requires extensive analysis from the movement model perspective. It is shown that conventional methods of modelling fish behaviour are inconvenient. This can be seen where specific motion could be modelled, but it will disregard other factors; for instance, the study on tracking the movement, the speed and number of bait or meal approaches has been utilized [9–11]. Nonetheless, the overall comprehension of feeding behaviour could not be justified as a whole, specifically in time-series responses. These issues have been eradicated by applying AI method explicitly using machine learning (ML) techniques [12, 13]. It is a method that learns from experience; in this case, the data sets are monitored before the prediction of behaviour provided the information collected are well organized accordingly. For example, the indication of fish behaviour must show evidence of the changing phases between hungry and satiated phases by which several monitoring methods have been applied such as sonar sensor, image processing, mechanical induced baits and electrical triggering sensors [14, 15]. Once the encompassing information has been collected, the ML techniques through a comprehensive pipeline are being analysed from the data collected, identifying the event occurrence, clustering the responses, extraction and selection the significant features, comparing with other classifiers and lastly optimizing the parameters should elevate the accuracy in identifying the hunger state.

The overall rationalization of the methods has been briefly described. The main objective, in general, is to identify the state of hunger in fish using automated demand feeder through imageries from fish movements. This is further analysed using the ML technique that developed through a series of methodological data analytics approach to acquire the best paradigm that emulates the feeding pattern. In the next section, the elaboration of fish hunger behaviour is discussed; this will highlight the significant parameters that represent the motion of fish regarding hunger routine.

1.2 Fish Hunger Behaviour

There have been several studies on extracting the information on the fish movement that relates between fish movement and their ecological environment as to what the hunger behaviour meant [9, 16, 17]. The biological sense towards hunger motivation is the main factors that contribute to their instincts in searching or foraging for food as this generally Zeitgeber time or the circadian rhythm. This pattern has somewhat related to other behavioural changes to other mammal anatomies, for instance, the sleeping and awakening juncture [18]. Either sleeping or feeding behaviour changes, it is said to persist in a 24 h routine and repeatedly for the subsequent day where some scientist has termed this as a free-running state [19].

With the appropriate manner in monitoring the feeding schedule, a model based on the movement can be measured provided the significant elements such as the time of feeding, amount of food to be dispersed and the intervals between each food droppings to be gathered. The importance of providing accurate determinants would lead to a better growth rate of the fish to be reared [20]. Similarly, it is vital that the allowable volume given to the school of fish does not surpass their consuming capabilities, as this might impair feeding efficiency that will cause overfeeding and increase the wastages of food [9]. By exploiting these parameters of feeding time and food volume, a rationalization of fish appetite can be acquired, which will provide the state between hungry and satiated phases.

Commonly, the activity of fish movement either individually or in a group reduces as it reaches satiated state compared to the predicament of being famished where the motion suggests to varied more [21, 22]. Hence, by capitalizing this rudiment of ever-changing motion, the relation can be considered as the circadian cycle of feeding pattern [23]. In another study, the changes in illumination level in the environment do influence the motion of fish incongruent with hunger state [19]. The perception is that the fish are naturally behaving accordingly between day and night conditions; therefore, by changing the light status would manipulate the hunger behaviour as well. In a similar note, other studies have developed the sense of activity shifts from other components namely by the social behaviour, sensory cues, species type and environmental changes as the time of day as mentioned earlier.

The extraction of significant features, in this case, the element that alters the fish behaviour, must consider not just a single parameter. However, one should include different criteria where the pattern can be justified thoroughly. This will induce multi-oscillator where the distinctive set of important factors can demonstrate the cause of effects of the identifying study. This statement was gathered from a survey in which highlighted that the relation between influences is not apparent with regard between feeding time and frequency; however, other parameters might provide the endogenous nature of the fish to imitate the circadian rhythm [24]. Besides, the authors underlined that a systematic approach on establishing an experiment data collection set-up is decisive for any relevant studies on the variables such as the temperature, species and the light conditions. Figure 1.1 illustrates some of the cases

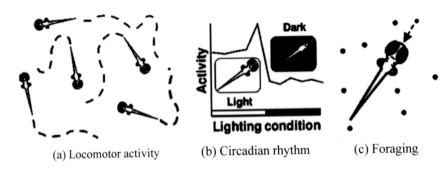

(a) Locomotor activity   (b) Circadian rhythm   (c) Foraging

Fig. 1.1 Swimming behaviour patterns

in which the behavioural changes to be determined before constitute a monitoring device for behavioural changes [25].

The locomotor activity, as in Fig. 1.1a, depicts the movement of the fish in groups. The tracking motion in this example shows the individual path which commonly implemented. The circadian rhythm patterns, on the other hand, show causes by manipulating the dark and light condition as in Fig. 1.1b that influences the loco-motor activities, likewise during scavenging of food, the motion and direction of the individual will be directed towards the food pellet as in Fig. 1.1c. The significance of understanding the feeding behaviour in terms of motion and movement will provide better insight on the insistence of the fish. This is crucial in aquaculture industries where researchers have to establish that the growth rate of cultured fish is highly correlated to the efficiency of the feeding routine [26].

To summarize this section, the distinctive parameters from the fish behaviour have their significance towards the rhythmic feeding patter of hunger. For example, the temperature, speed of the fish's motion, time of food supply and the intervals, environmental changes and water condition would alter the hunger state of the fish. Disregard the subjects monitored either in groups or individually, and the activi-ties suggest more rapidly as the fish ten to become more famish. By employing the relationship between hunger state and the significant behavioural changes, this study resembles the rearing application for providing constructive evident from the fish hunger state. The following section will discuss the method, specifically the image processing technique, in which this study utilizes in capturing the significant parameters mentioned.

1.3   Image Processing Parameter

The procedural approach in this study is to use computer vision as the monitoring device in gathering fish behavioural changes along with the demand feeder system as the food dispenser for feeding rhythmic recognition. This section will elaborate from the previous fish behaviour parameter with the concept of utilizing image processing as the data extraction for hunger behaviour prediction. It was seen in recent years that camera vision had been adapted in the aquaculture field to capture the movement of fish for studying the swimming patterns [13, 27, 28]. Research has attempted by implanting a device on the subject to examine the starvation status to evaluate the muscle activity of the fish and thus to specify the appetite status [29]. This may involve greater costs if the research involves monitoring larger quantities of fish, so a more practical alternative is needed to solve this. One of the main reasons for using image processing compared to other techniques such as sonar and multiple sensors is the coverage area of the swimming environment which can be captured in a wider range, and the precision in tracking the motion is better. For instances, studies have used pixel calculation as the tracking identifier from the fish movement by defining the centroid or centre of gravity (COG) of the individual or in a group [13, 30].

Other factors that influence the behaviour were being studied where the correlation between speed and satiation rate is observed from the volume and density of the food supplied [10]. The method used is by analysing each of the frames from the video whereby the speed of the fish changes as the number of attacks or, in this case, the food scavenging rate. The distance between the fish and food area was investigated as this is to determine the satiation level to justified had the fish being fed. It was found that as fish getting hungrier, the number of attacks increases in contrast to the density of the food where the density of food is inversely proportional to the speed of foraging. Even though the circadian pattern was not thoroughly elaborated, however, it can be presumed that the information has valuable insight on the movement activities towards the hunger state of the fish. The definitive of a constructive experimental set-up is the ability of the device to apprehend the distinctive parameters that contribute to the finding of feeding routine pattern. A study was done to find the correlation between the speed in straight motion and turning and the distances among the individuals of the school [31]. Although it was found that fish scavenges more in unfamiliar search area when they tend to get hungrier, the result did not explicate in a continuous feeding reading, whereas as time goes by the motion of fish reduces as the feeding reaches a fully fed state. Nonetheless, the justification has somewhat told essential input on fish being in a group setting, in which the response between individuals has higher tendency rather than acting on as a whole party. Another study has conducted an examination on detecting the impact of boldness and the aggressiveness on starvation level with different light intensity environment [11]. This is considered as a good example to investigate the hunger level. However, the experiments were carried out without the use of an automated monitoring device, this in turn, limits the data extraction and evaluation should longer period is required. Nonetheless, they have concluded that gender differences do not influence much on the rate of aggressive or boldness as the light condition changes, but the female subjects will become more aggressive as they become hungrier. Similarly, an investigation was done to find whether fish would overcome their personal trait of bold and shyness when compared separately form the group of fish [22]. The discrepancy between the two is that the animals that fear to scavenge in the dangerous eating region are energetic and bold when the animals spend longer outside the deck or safe area. It was noted, on the other hand, that the shyer fish will become more involved or imitate the bolder associate because of food scavenging when they are not fed. Nevertheless, the repeatability of eating habits was deficient in view of the detection of recurrence activities on the hungry fish, because the technique was performed in two hours which is difficult to predict the subsequent perioding feeding pattern. On the conceptual design in underlining the motion analysis or emphasizing movement assessment of eating behaviour, it has shown that by integrating image processing techniques and geometrical calculation, it could track the motion of the fish [13]. Admirably, the study has demonstrated a test configuration that manages the relationship food wastages with the self-inflicting feeder device. The study shows that the combination of detectors and feed equipment can confirm the feeding trend of group fish using image processing.

The basics for the extraction of important fishery parameters have been set in this chapter for assessments of multiple techniques of inclusion. The requirement to determine the stage of hunger and satiation is described so that the results are referred to. This emphasizes the need to develop a guideline on the perception of hunger. The support instruments such as camera vision and imaging handling methods also comprehensively increase the surveillance accuracy of eating trends of the fish. The images obtained must be justified, as this is used to extract the important monitoring characteristics. An embedded scheme, which meets these demands, could, therefore, be reinforced where it was jointly created to provide data about the behaviour of fish-level starvation.

Nevertheless, in order to classify and forecast the hunger behaviour, the link between eating and circadian trends in time series was not detailed because of various variables, including daytime, social interactions, eating mechanisms and environmental indicators. The model for providing the conclusions of the projection of the hunger condition was not deliberated because the behaviour of fish hunger could not be parametrically modelled, but only in a corresponding way. ML includes elements of behaviour modelling because it needs data sets to be used as a template to predict future outputs. The next section will, therefore, explore the ML techniques used to apply the methods in identifying fish behaviour to provide a clearer understanding of the approaches suggested for this research.

1.4 Machine Learning in Fish Behaviour

Extracting the significant characteristics is essential in order to provide data to classify the training model for computer vision algorithms because they could recognize the fish motion. In this section, ML techniques adopted in fish behaviour will be discussed to help the approaches used in this study in constructing fish hunger behaviour prediction according to classifier designs, classification methods, extract characteristics and analyses.

The use of artificial intelligence (AI) in the identification of fish behaviour has been a trend lately as it makes behavioural modelling less complex and therefore produces considerable predictive designs from nonlinear information [32–34]. Most assessments have a notion to describe movement monitoring but do not take into account modifications in the hunger state [13, 35]. The feature extraction method must be the corresponding fish location and moment prior to the evaluation of hunger behaviour and must be precise for the achievement of desired outcomes. These parameters are used to identify the techniques of identifying and classifying fish to follow their eating behaviour. A thorough assessment is required to determine the finest parameter for the precision of classification.

The partitioning of behaviours needs a clustering stage in which research has shown that school fish are detected using k-means techniques [36]. In addition, a clustering analysis was carried out in another research to determine the appropriate cluster number for the distribution of fish species within the ecological community [37]. The interpretation of fish behaviour from characteristics often involves an elevated input level, where information streaming from video transcends at a large pace and transforms abandon the quantity of information to be classified. The choice of characteristics could, therefore, decrease the number of larger features by analysing the box plot analysis or decreasing dimensionality by using principal component analysis (PCA) [21, 38].

In separate types of case studies on fish, an ML algorithm has been exhibited as a credible means to identify the behaviour of fisheries. k-NN is another technique that was successful in the past instance in recognizing hunger of the fish classification problem [39]. The support vector machine (SVM) has also shown usability in the classification of variables in an effective behaviour representative [34, 40]. Different research has used SVM or other classifications namely discriminant analysis (DA), logistic regression and artificial neural network (ANN) techniques to promote their theory and are considered to be adequate [41–43]. Different ML designs are peculiar and sometimes work better for varying information characteristics. It is, therefore, necessary to assess the efficiency of distinct designs in the classification of hunger.

References

1. Fisheries Department of Malaysia (2016) Landings of marine fish by tonnage class and fishing gear group
2. de Mattos BO, Filho ECTN, Barreto KA, Braga LGT, Fortes-Silva R (2016) Self-feeder systems and infrared sensors to evaluate the daily feeding and locomotor rhythms of Pirarucu (*Arapaima gigas*) cultivated in outdoor tanks. Aquaculture 457:118–123. https://doi.org/10.1016/J.AQUACULTURE.2016.02.026
3. Cho CY (1992) Feeding systems for rainbow trout and other salmonids with reference to current estimates of energy and protein requirements. Aquaculture 100:107–123. https://doi.org/10.1016/0044-8486(92)90353-M
4. Parra L, Sendra S, García L, Lloret J (2018) Design and deployment of low-cost sensors for monitoring the water quality and fish behavior in aquaculture tanks during the feeding process. Sensors 18:750. https://doi.org/10.3390/s18030750
5. Swann L (1997) A fish farmer's guide to understanding water quality. In: Aquaculture extension. Aquaculture Extension, Illinois-Indiana Sea Grant Program, p AS-503-511
6. Pillay TVR, Kutty MN (2005) Aquaculture: principles and practices. Blackwell Publishing

7. Mukai Y, Tan NH, Khairulanwar M, Chung R, Liau F (2016) Demand feeding system using an infrared light sensor for brown-marbled grouper juveniles, *Epinephelus fuscoguttatus*. Sains Malays 45:729–733

8. Volkoff H, Peter RE (2006) Feeding behavior of fish and its control. Zebrafish 3:131–140. https://doi.org/10.1089/zeb.2006.3.131

9. Alós J, Martorell-Barceló M, Campos-Candela A (2017) Repeatability of circadian behavioural variation revealed in free-ranging marine fish. R Soc Open Sci 4:160791. https://doi.org/10.1098/rsos.160791

10. Priyadarshana T, Asaeda T, Manatunge J (2006) Hunger-induced foraging behavior of two cyprinid fish: *Pseudorasbora parva* and *Rasbora daniconius*. Hydrobiologia 568:341–352. https://doi.org/10.1007/s10750-006-0201-5

11. Ariyomo TO, Watt PJ (2015) Effect of hunger level and time of day on boldness and aggression in the zebrafish *Danio rerio*. J Fish Biol 86:1852–1859. https://doi.org/10.1111/jfb.12674

12. Allken V, Handegard NO, Rosen S, Schreyeck T, Mahiout T, Malde K (2019) Fish species identification using a convolutional neural network trained on synthetic data. ICES J Mar Sci 76:342–349. https://doi.org/10.1093/icesjms/fsy147

13. Zhou C, Zhang B, Lin K, Xu D, Chen C, Yang X, Sun C (2017) Near-infrared imaging to quantify the feeding behavior of fish in aquaculture. Comput Electron Agric 135:233–241. https://doi.org/10.1016/j.compag.2017.02.013

14. Parra L, García L, Sendra S, Lloret J (2018) The use of sensors for monitoring the feeding process and adjusting the feed supply velocity in fish farms. J Sens 2018:1–14. https://doi.org/10.1155/2018/1060987

15. Ogunlela AO, Adebayo AA (2016) Development and performance evaluation of an automatic fish feeder. J Aquac Res Dev 07:1–4. https://doi.org/10.4172/2155-9546.1000407

16. Braithwaite VA, Rosenthal GG, Lobel PS (2006) Circadian rhythms in fish. In: Behaviour and physiology of fish, vol 24, pp 197–238. https://doi.org/10.1016/s1546-5098(05)24006-2

17. Cavallari N, Frigato E, Vallone D, Fröhlich N, Lopez-Olmeda JF, Foà A, Berti R, Sánchez-Vázquez FJ, Bertolucci C, Foulkes NS (2011) A blind circadian clock in cavefish reveals that opsins mediate peripheral clock photoreception. PLoS Biol 9:e1001142. https://doi.org/10.1371/journal.pbio.1001142

18. Beale A, Guibal C, Tamai TK, Klotz L, Cowen S, Peyric E, Reynoso VH, Yamamoto Y, Whitmore D (2013) Circadian rhythms in Mexican blind cavefish *Astyanax mexicanus* in the lab and in the field. Nat Commun 4:2769. https://doi.org/10.1038/ncomms3769

19. Sanchez-Vázquez FJ, Madrid JA, Zamora S (1995) Circadian rhythms of feeding activity in sea bass, *Dicentrarchus labrax* L.: dual phasing capacity of diel demand-feeding pattern. J Biol Rhythms 10:256–266. Also: https://doi.org/10.1177/074873049501000308

20. Rahman MM, Nagelkerke LAJ, Verdegem MCJ, Wahab MA, Verreth JAJ (2008) Relationships among water quality, food resources, fish diet and fish growth in polyculture ponds: a multivariate approach. Aquaculture 275:108–115. https://doi.org/10.1016/J.AQUACULTURE.2008.01.027

21. Kennedy J, Jónsson SÞ, Ólafsson HG, Kasper JM (2016) Observations of vertical movements and depth distribution of migrating female lumpfish (*Cyclopterus lumpus*) in Iceland from data storage tags and trawl surveys. ICES J Mar Sci J du Cons 73:1160–1169. https://doi.org/10.1093/icesjms/fsv244

22. Nakayama S, Johnstone RA, Manica A (2012) Temperament and hunger interact to determine the emergence of leaders in pairs of foraging fish. PLoS ONE 7:e43747. https://doi.org/10.1371/journal.pone.0043747

23. Chapman BB, Morrell LJ, Krause J (2010) Unpredictability in food supply during early life influences boldness in fish. Behav Ecol 21:501–506

24. Boujard T, Leatherland JF (1992) Circadian rhythms and feeding time in fishes. Environ Biol Fishes 35:109–131. https://doi.org/10.1007/BF00002186

25. Mora-Zamorano FX, Klingler R, Basu N, Head J, Murphy CA, Binkowski FP, Larson JK, Carvan MJ (2017) Developmental methylmercury exposure affects swimming behavior and foraging efficiency of yellow perch (*Perca flavescens*) larvae. https://doi.org/10.1021/acsomega.7b00227

26. Harpaz S, Hakim Y, Barki A, Karplus I, Slosman T, Tufan Eroldogan O (2005) Effects of different feeding levels during day and/or night on growth and brush-border enzyme activity in juvenile *Lates calcarifer* reared in freshwater re-circulating tanks. Aquaculture 248:325–335. https://doi.org/10.1016/j.aquaculture.2005.04.033

27. Thida M, Eng H, Chew BF (2009) Automatic analysis of fish behaviors and abnormality detection. PROC IAPR Mach Vis Appl 8–18

28. Volpato GL, Bovi TS, de Freitas RHA, da Silva DF, Delicio HC, Giaquinto PC, Barreto RE (2013) Red light stimulates feeding motivation in fish but does not improve growth. PLoS ONE 8:e59134. https://doi.org/10.1371/journal.pone.0059134

29. Cubitt KF, Williams HT, Rowsell D, McFarlane WJ, Gosine RG, Butterworth KG, McKinley RS (2008) Development of an intelligent reasoning system to distinguish hunger states in rainbow trout (*Oncorhynchus mykiss*). Comput Electron Agric 62:29–34. https://doi.org/10.1016/j.compag.2007.08.010

30. Spampinato C, Chen-Burger Y-H, Nadarajan G, Fisher R (2008) Detecting, tracking and counting fish in low quality unconstrained underwater videos, pp 514–519

31. Hansen MJ, Schaerf TM, Ward AJW (2015) The effect of hunger on the exploratory behaviour of shoals of mosquito fish *Gambusia holbrooki*. Behaviour 152:1659–1677. https://doi.org/10.1163/1568539X-00003298

32. Hasija S, Buragohain MJ, Indu S (2017) Fish species classification using graph embedding discriminant analysis. In: 2017 international conference on machine vision and information technology (CMVIT 2017): 17–19 Feb 2017, Singapore. IEEE, pp 81–86

33. Iswari NMS, Wella R (2017) Fish freshness classification method based on fish image using k-Nearest Neighbor. In: 2017 4th international conference on new media studies (CONMEDIA): 8–10 Nov 2017, Yogyakarta, Indonesia. IEEE, pp 87–91

34. Sudana M, Nalluri R, Saisujana T, Reddy KH, Swaminathan V (2017) An efficient feature selection using artificial fish swarm optimization and SVM classifier. In: 2017 international conference on networks and advances in computational technologies (NetACT). IEEE, pp 412–416

35. Xu Z, Cheng XE (2017) Zebrafish tracking using convolutional neural networks. Sci Rep 7:42815. https://doi.org/10.1038/srep42815

36. Buelens B, Pauly T, Williams R, Sale A (2009) Kernel methods for the detection and classification of fish schools in single-beam and multibeam acoustic data. ICES J Mar Sci 66:1130–1135. https://doi.org/10.1093/icesjms/fsp004

37. Jackson DA, Walker SC, Poos MS (2010) Cluster analysis of fish community data: "new" tools for determining meaningful groupings of sites and species assemblages. Am Fish Soc Symp 73:503–527

38. Wishkerman A, Boglino A, Darias MJ, Andree KB, Estévez A, Gisbert E (2016) Image analysis-based classification of pigmentation patterns in fish: a case study of pseudo-albinism in Senegalese sole. Aquac 464:303–308. https://doi.org/10.1016/J.AQUACULTURE.2016.06.040

39. Razman MAM, Susto GA, Cenedese A, Abdul Majeed APP, Musa RM, Abdul Ghani AS, Adnan FA, Ismail KM, Taha Z, Mukai Y (2019) Hunger classification of *Lates calcarifer* by means of an automated feeder and image processing. Comput Electron Agric 163:104883. https://doi.org/10.1016/J.COMPAG.2019.104883

40. Bermejo S (2014) The benefits of using otolith weight in statistical fish age classification: a case study of Atlantic cod species. Comput Electron Agric 107:1–7. https://doi.org/10.1016/J.COMPAG.2014.06.001

41. Cortes C, Vapnik V (1995) support vector networks. Mach Learn 20:273–297. https://doi.org/10.1023/A:1022627411411

42. Dutta MK, Sengar N, Kamble N, Banerjee K, Minhas N, Sarkar B (2016) Image processing based technique for classification of fish quality after cypermethrin exposure. LWT - Food Sci Technol 68:408–417. https://doi.org/10.1016/J.LWT.2015.11.059

43. Ogunlana SO, Olabode O, Oluwadare SAA, Iwasokun GB (2015) Fish classification using support vector machine. Afr J Comput ICT Afr J Comput ICT Ref Format Afr J Comp ICTs 8:75–82

Chapter 2
Monitoring and Feeding Integration of Demand Feeder Systems

Abstract This chapter highlights the findings of the developmental monitoring systems for swimming pattern or motion analysis with regard to feeding behaviour. A benchmark for examining the framework on how scientists control fish in animal variable function factors was gathered and referred to gauge the adequate design in constructing a viable device. The validation of image processing and automated demand feeder to determine the results will also be considered, as a validation aspect between the system of tracking and the behaviour of the *Lates calcarifer* where the pixel intensity will be extracted as the features. The results of this chapter will enable the reader on the development of an integrated feeder scheme that consolidates surveillance scheme to identify the feeding behaviour and relation towards the specific growth rate (SGR).

Keywords Image processing · Automated demand feeder · *Lates calcarifer* · Pixel intensity · Specific growth rate

2.1 Overview

The findings on the development of fish hunger monitoring systems that validate swimming habits are presented in this section. The analysis will focus on identification ideas, techniques for movement monitoring, type of species used, the distribution of supplies and general scheme set-up that implement with image processing and automated feeder system. A benchmark has been examined for how scientists track fish from animal variable function factors [1]. In which the industrial technology concepts implemented in other sections and the principal difficulties to introduce them into manufacturing aquaculture have been categorized namely sonar, acoustic telemetry tags, passive acoustics remote controlled and computer vision. This study will be focusing on the variables reading of behavioural characteristics of swimming patterns using camera as the monitoring device. However, the factors of observing the motion will be targeted towards hunger-induced activation as some researcher has observed the correlation between speed and hunger rate by altering the feeding volume [2]. The amount of approaches and satiations' rates, where the density of two meal densities was calculated by comparing images that were captured using the

M. A. Mohd Razman et al., *Machine Learning in Aquaculture*,
SpringerBriefs in Applied Sciences and Technology,
https://doi.org/10.1007/978-981-15-2237-6_2

video system, consists of two cameras for top and side views. Another research had created an experimentation configuration to examine these parameters in developing a notion of the subject's velocity, turning speed and ranges between subjects [3]. There were two barriers in the way of the locomotive activity, where the motion was captured using a camera. The tests were conducted through the release of four fish for two distinct communities of starving and satiated fish in each gender. They found out that relative to satiated fish, hungrier subjects showed a greater rate of motion and rotation. These are some of the methods used to gather information towards the motion activity. However, the integration with an automated feeder is hardly seen. Generally, the feeding that was done was given manually instead of triggering certain sensors. On the contrary, automated feeder mostly emphasizes on distinctive study to examine the specific growth rate (SGR) [4, 5]. An example of applying demand feeder is by suspending a sensor onto the water as a targeting reference for the fish [6]. The fish school triggers the sensor due to hunger as they upsurge towards the dispensing area, and the feeder will start to drop the food. The subjects have shown to increase the frequency of sensor activation while in starvation for food and gradually subsidizes and when it shows to satiate. Another study has demonstrated similar concepts on integrating monitoring sensors and demand feeder to analyse the amount taken from the school of fish and should there be any wastages occur [7]. The food supply mechanism can improve fish growth, feed efficiency, lower feed losses and water pollution, because fish have adapted to the consumption of feed. However, the variations of fish growth have proved to be diverse as a certain individual is involved in the designated area of trigger and food dropping sections [8, 9]. Therefore, this study implements the combinational concept of automated demand feeder and image processing to predict the behavioural understanding of hunger states in a school size form. This will provide a justification for monitoring the entire school fish and validate the feeding frequencies as the circadian rhythm can be recorded. The next section will elaborate on the experimental set-up where the integration between computer vision and demand feeder will be explained before succeeding in data processing and analytics in later chapters.

2.2   Fish Monitoring Set-Up

In developing the proposed monitoring set-up, the prospects of a fish hunger classification were taken from the literature, and several problems that should be addressed before it fits into the projected model have been considered. The variables leading to impulsive operations, in which the circadian rhythm could be achieved, are identified since the automatic feeder reinforces the data with a camera vision surveillance system. A synchronization of locomotive information and the listing of feeding moments would lead to a circadian rhythm. The suitable demand feeder was developed in order to support the aim of this study by showing the practicality and effectiveness of the food provision to the group of fish. Hence, the following setup of image processing will be elaborated to supplement the device and demonstrate the integration between the motion and feeding patterns. This will be resulted in the SGR to validate the performance of the proposed method, especially on the demand feeder. This section will start by describing the automated demand feeder, on the implementation approaches, followed by the image computer vision set-up.

2.2.1 Experimental Materials and Set-Up

Figure 2.1 illustrates the layout and advancement of the automatic feeder. The mechanical design and positioning of the electronic systems including the microcontroller, motor and infrared (IR) sensors are demonstrated as the main components for the system as in Fig. 2.1a. The pellets are inside the pellet container that can refill that can last several weeks of supply. The rotary feeder is where, when the motor rotates, a fragment of a pallet is dispersed when it is triggered by the IR sensor at the bottom of the feeder. The IR sensor is dipped in the water where fish can activate the sensor as a point of initiation for demand. The stepper motor is located below the container as it is connected to the rotary feeder, as shown in the rear view in Fig. 2.2a. On the side of the stepper motor is the microcontroller where it is embedded with all the sensors and actuators, and the entire automatic feed system is enclosed in the body of the container.

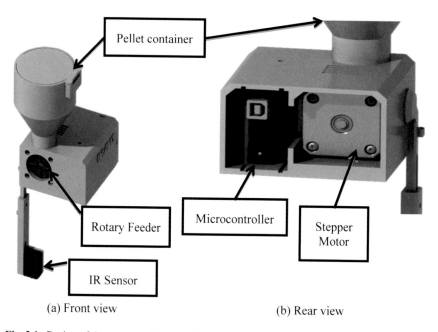

(a) Front view   (b) Rear view

Fig. 2.1 Design of the automated demand feeder

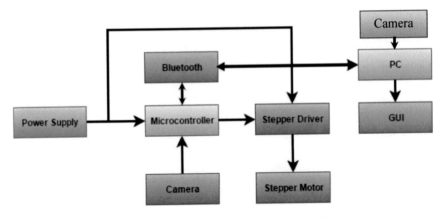

Fig. 2.2 Integration of the automated feeder system with IR sensor and camera

Table 2.1 Component for demand feeder device with the image processing system

Materials	Specification and function
Power supply	DC Adapter 12 V/2 A is the main power supply for the fish feeder
Microcontroller	Bluno M3 to receive input from IR sensor when the fish are detected and send signals to stepper driver to drive the stepper motor to feed the fish
Stepper driver	A4988 stepper driver to control stepper motor direction and rotation
Stepper motor	NEMA 17 stepper motor actuator which drives the feeding mechanism
Infrared sensor	The IR sensor is the trigger point for the fish as it can be detected underwater, Panasonic CX-422
HD Wi-Fi camera (DCS-936L)	720p resolution or 1280 × 720 pixel size and could capture up to 30 frames per second (fps) Record the total times of the feeder machine being triggered
Feeder casing	Acrylonitrile butadiene styrene (ABS) 3D printed material
Personal computer	Intel® Core™ i7-7500 at 2.70 GHz processor with an installed memory of 8 GB RAM
Aquarium	92 cm × 46 cm × 46 cm with 130 L of water filled
Lates calcarifer (Barramundi) fish species	20 juvenile fish as the test subjects The average body weight is 0.58 ± 0.05 g, and the average total length is 33.2 ± 0.08 mm
Food pellet krill and shrimp	10–15 pellets with every drop an approximate size of 3 mm in diameter. 42% protein contents

The materials and functions of each component are described as in Table 2.1 where the process flow chart of the system is depicted as in Fig. 2.2.

The list of three main sections that is mechanical, electrical and electronics followed by the data acquisition components can be seen in Table 2.1. The pipeline of working principles of the system follows the structural flow chart as depicted in Fig. 2.2 where it starts by supplying the main supply of DC adapter with a 12 V and 2 A. It is supplied to both the microcontroller and the stepper driver that support the rotational torque of the motor. The main source of the signal received from the trigger sensor is the microcontroller which provides fish with food by rotating the step motor once activated. The IR sensor is embedded in the microcontroller where the supplied voltage is taken out from the microcontroller board. After assembly, the housing and parts were printed using ABS material for the whole model with a total of 356 g. The sensitiveness of fish detection was adjusted to accommodate the fish in a small cluster as specified in this study, only then the rotating feeder is activated. Some fish have the capacity to see infrared light, but the effects are not biological and only stimulate the motivation of fish when exposed in a red light environment [4]. The species of *Lates calcarifer* or commonly known as Asian barramundi are used for the test subjects which naturally adapted to the environment of laboratory condition as it is accustomed to the Asian region climate. In order to record the behaviour of a fish, the demand feeder using an image processing system then sends the captured images to the PC. The camera used is sufficient to capture the entire field of view of the aquarium with a 100° of angle for horizontal, 54° of vertical width and 120° in diagonal. The images' behaviour of the fish was analysed by the software RoboRealm, which selected and extracted the characteristics of the motion, e.g., the parameters for the fish group and the activities of the fish movement. The integration of the entire system can be illustrated as in Fig. 2.3.

A laboratory at the International Islamic University Malaysia (IIUM) was the location on collecting the data concerning the fish. Automated feeder development has been developed in advance at the Innovative, Manufacturing, Mechatronics and Sports (IMAMS) laboratory, University of Malaysia Pahang (UMP), to ensure that the feeder is used for the data collection of fish. The cooperation of both parties resulted in the introduction of a patent no.: PI 2017703506, allowing inventors to use the experimentation set-up [10].

The experimental set-up consists of a workstation with the fish movement monitoring, and the aquarium section is illustrated as in Fig. 2.4. Black cloth covers the aquarium section to avoid interruptions that might result from human interactions as depicted in Fig. 2.4a, and the aquarium sections are placed as in Fig. 2.4b. The demand feeder is placed on top of the aquarium as shown in Fig. 2.5. The camera as illustrated in Fig. 2.3 is placed on the front side of the aquarium that captures the entire field area of fish movement which can be seen in Fig. 2.6.

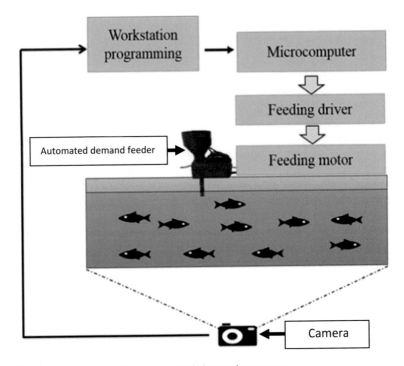

Fig. 2.3 Camera and demand feeder schematic integration system

(a) Front view Setup   (b) Aquarium section

Fig. 2.4 Experimental set-up

Fig. 2.5 Automated feeder system model set-up

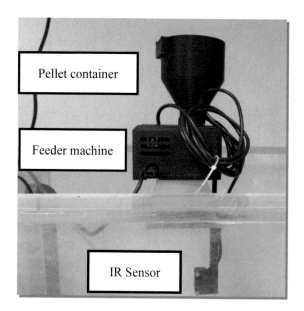

Fig. 2.6 Camera position capturing fish tank front view set-up

2.2.2 Image Processing Data Extraction

The purpose of this section is to determine the features to be taken from images taken in the integration between the automated feeder system and the video camera. The process will start with a real-time collection of the video and ensure that RoboRealm is compatible with the video format. After transmitting the videos to the software, the functions installed in the software can be used to collect the required picture, which only separates the pattern of fish swimming. Once the video is gathered, the image processing techniques are applied using the software to segregate the fish from other environment figures where cropping is done. Figure 2.7 illustrates the resulting

(a) Image from camera   (b) Image after cropping

Fig. 2.7 Image of cropping

image crop result. Figure 2.7a shows that the filter system, and the bubbles are not included in the cropped images as resulted in Fig. 2.7b.

The fish and background photographs are isolated by means of a contrast module, in which the images are sharpened, when the contrast adjustments occur, as the bias set of black and white pixels are induced as Fig. 2.8a shows the contrast picture. The threshold module for discarding parts of the images which fall within a specific intensity range, which can be adjusted between 0 and 255, has been set, and the outcome is shown in Fig. 2.8b. The resulted figures on separating the background from the original images with the group of fish are listed in Table 2.2. The next

(a) Image after contrast   (b) Image after threshold

Fig. 2.8 Image of contrast and thresholding

Table 2.2 Parameters set-up on the image processing

Steps	Function	Description
1	Capture	Transferring the video into the software in a compatible format (.avi)
2	Crop	X start = 158, X end = 892, Y start = 361, Y end = 662
3	Contrast	60 contrast value
4	Threshold	0–124 intensity
5	COG	Extracting features from the images

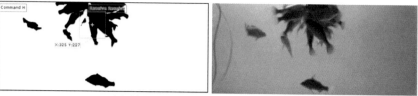

(a) Fishes during triggering the sensor

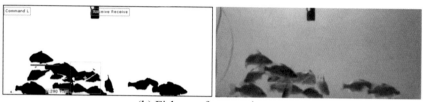

(b) Fishes at free running sate

Fig. 2.9 Image at each frame processing parameter

section will elaborate on the relationship between the movements throughout the experiments and validate the demand feeder according to the SGR.

2.3 Fish Growth

SGR has been proven to display the effectiveness of a certain study or to evaluate the performance of a particular device, whether it has reached a feasible result as expected [4, 5]. It commonly compared two different approaches; for instance in this study to evaluate the efficacy of the automated demand feeder, it is gauge with another set of manually induced feeding. The difference between the original and final weight as well as length is measured and split by the duration of the experiment in order to establish the growth in weight and length between the two groups. The following equations can be deduced [8].

$$SGR_{TL} = \frac{TL_f - TL_i}{\text{Total Rear Days}} \times 100 \tag{2.1}$$

$$SGR_{BW} = \frac{BW_f - BW_i}{\text{Total Rear Days}} \times 100 \tag{2.2}$$

The notation of SGR_{TL} represents the specific growth rate on the total length, TL, and SGR_{BW} is the specific growth rate on the body weight, BW. The final and initial measurements are subscripted as f and i for TL and BW, respectively. Prior to validating the effectiveness of the fish monitoring system, the motion that was captured has been transmitted according to the experimental methodologies that relate with hunger which is the activation of the sensor and the time spent before

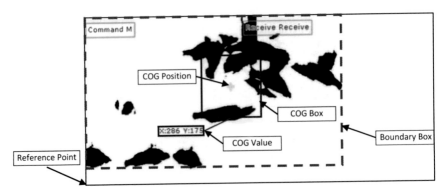

Fig. 2.10 Fish position **a** during activation of sensor **b** free-running state

stop dispersing the food supply. The time of the meal is entered into a database to identify the patterns of starvation. Depending on how often a fish triggers the sensor, the cycle is repeated. The collection time was taken for 13 h from 8:00 a.m. to 9:00 p.m., and the images of the fishes' activity can be seen in Fig. 2.10 that display the moment when the group trying to activate the sensor and at free-running state and kept at 14 days period for comparing automated and manual feeding (Fig. 2.9).

After the images have met the required condition, motion analysis is easier to implement by manipulating changes in the pixel. Adjustment of the RoboRealm software on the centre of gravity (COG) interface enables extraction of variables from the video that is captured where the fundamentals of COG or sometimes called the centroid are derived as follows [11]:

$$COG\_X := \frac{\sum_0^n I_n x_n}{\sum_0^n I_n} \tag{2.3}$$

$$COG\_Y := \frac{\sum_0^n I_n y_n}{\sum_0^n I_n} \tag{2.4}$$

$$I_n := \frac{R_n + G_n + B_n}{3} \tag{2.5}$$

The pixels as expressed in the equations show that COG_X is the centre of gravity on the x-axis, where COG_Y is the COG for y-axis. The pixels' intensity, I_n, is the summation throughout the updated R G and B denoting for red, green and blue pixels, respectively. The denotation of x_n and y_n is the position of the pixel on x-axis and y-axis, respectively. The mentioned pixels are based on the picture of each video frame where Fig. 2.9 illustrates the pixels change according to the frame at the given time. Henceforth, the collective parameters that were collected are listed as in Table 2.3.

The chosen parameters were based on the software's ability and hunger relationship as this is analyzed in later section feature extraction and selection. It is crucial to determine the time when fish activates the sensor when data from the automated

Table 2.3 Parameters descriptions gathered from computer vision

Parameters	Description
COG_BOX_SIZE	Width or height of the square bounding box
COG_X	X-coordinate of the COG
COG_Y	Y-coordinate of the COG
COG_AREA	Number of non-black pixels that contributed to the COG calculation
DENSITY	A measure of how densely packed the pixels are that contributed to the COG calculation
MOVE_PIXEL	Number of pixels that were deemed as changed in comparison to the background image
MOVE_SUM	Summed amount of pixel differences

feeder are received. Therefore, the parameters, as listed in the table above, are transferred in real time to gather the timestamp into Excel file. In addition, the movement is captured in account of a group based rather than individually.

2.4 Results and Discussion

In this section, the results of the study will be discussed accordingly. The results should make an important contribution to investigations of the fish hunger behaviour through the demonstration of the methods for experimentation in this study. The transmitted data are collected essentially via the system described in Fig. 2.2, as explained in the earlier section. The collected data can be displayed in Fig. 2.11 as

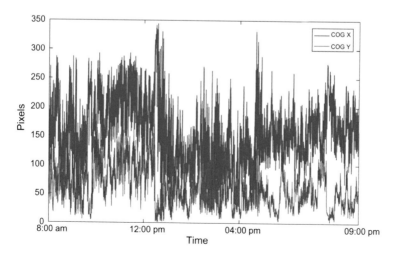

Fig. 2.11 Data collected for *COG_X* and *COG_Y*

some of the data which follow the image processed data structure based on Eqs. 2.3–2.5. The analysis will be done via MATLAB, where it was preceding gathered from the Excel file from the RoboRealm software. Another study has, for instance, demonstrated the concept of extracting the features according to the period of the fasted and fed fish which is similarly adopted in this study [12].

Figure 2.11 illustrates the examples of data of the motion on COG_X and COG_Y variables. In this case, the abnormal value of the pixel of fish can indicate that the audacity of all fish increases because most of them forage for food as this supported from a study that investigated the attitude that encourages the aggressiveness of fish during food deprivation [13]. To highlight the instances of food dropping and the relation between a variable, for example, COG_Y is shown in Fig. 2.12. The two groups that separate between 'Hungry' and 'Not Hungry' are manually categorized according to data given that the sensor was activated. It can be seen that during hungrier state, the fish location increases which means the school is trying to activate the sensor, whereas when they are not hungry the data have shown a calmer, or in this case, the fish stays relative lower than before. In general, a constant motion that shows the whole fish in a community has reached its full feeding phase, and this seems like a calmer attitude. The previous researcher also investigated this phenomenon of understanding the frequency of food prowling in sonar for observing the conduct of fish in close quarters and with the fish baits [14]. Overall, these findings indicate that the techniques specifically monitoring fish movement and food demands suggested in the earlier section are adequate, supported by other research which has been compiled with these researches, to obtain significant acquisition information from fish hunger behaviour.

As for the evaluation of performance for automated demand feeder, it was found that it produces a higher SGR for body weights with 6.90% as contrary to the manual feeding with 5.47%. Similarly, by using the manual feeding or time based, it suggests a lower rate of 1.93% for SGR of body length as compared to the automated demand feeder with 2.22%. The representation of the results can be seen in Fig. 2.13.

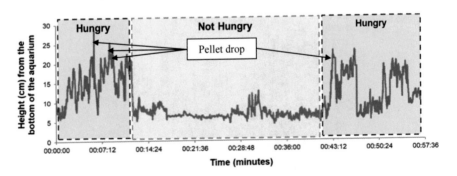

Fig. 2.12 Pellet dropping instances

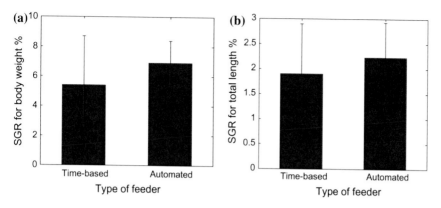

Fig. 2.13 SGR percentage between automated and time based **a** body weight **b** total length

The SGR for both the body weight and the length derived from the automatic feeder can be found to be 26% and 15% greater, respectively, compared to the time-based feeder. The integrated demand feeder device is therefore obviously superior to the standard feeder as the sensor to allow eating is activated on request. It would not only support the manufacturing and raising of these extremely requested fish in a laboratory setting but also could be applied and expanded on a wider scale as previously studied [15].

2.5 Summary

This chapter justified the idea that the device developed was supposed to rearing fish as it generates higher growth rates compared to time-consuming manual feeding. The integration between hunger-induced device and the image processing is validated as this system yield higher growth rate compared to the time-based feeding. The data collected through this research should warrant that the fish complies with hunger, as the sensor is triggered on request. The reasoning for classifying hunger behaviour is therefore rational, as the automatic demand feeder shows an increase in the development stage. The feed patterns can be realized by combining the requirements of the feed system and the fish monitoring system by collecting an accurate time of feeding time and using the information provided incomprehensible data. Hence, the classification of hunger behaviour can be realized which will be elaborated in the next chapter, mainly on the signal processing, clustering and feature extraction of the image processing data towards the hunger state behaviour.

References

1. Føre M, Frank K, Svendsen E, Alfredsen JA, Dempster T, Eguiraun H, Watson W, Stahl A, Sunde LM, Schellewald C, Skøien KR, Alver MO (2018) Precision fish farming: a new framework to improve production in aquaculture. Biosyst Eng 173:176–193. https://doi.org/10.1016/J.BIOSYSTEMSENG.2017.10.014
2. Priyadarshana T, Asaeda T, Manatunge J (2006) Hunger-induced foraging behavior of two cyprinid fish: *Pseudorasbora parva* and *Rasbora daniconius*. Hydrobiologia 568:341–352. https://doi.org/10.1007/s10750-006-0201-5
3. Hansen MJ, Schaerf TM, Ward AJW (2015) The effect of hunger on the exploratory behaviour of shoals of mosquito fish *Gambusia holbrooki*. Behaviour 152:1659–1677. https://doi.org/10.1163/1568539X-00003298
4. Volpato GL, Bovi TS, de Freitas RHA, da Silva DF, Delicio HC, Giaquinto PC, Barreto RE (2013) Red light stimulates feeding motivation in fish but does not improve growth. PLoS ONE 8:e59134. https://doi.org/10.1371/journal.pone.0059134
5. Benhaïm D, Akian DD, Ramos M, Ferrari S, Yao K, Bégout ML (2017) Self-feeding behaviour and personality traits in tilapia: a comparative study between *Oreochromis niloticus* and *Sarotherodon melanotheron*. Appl Anim Behav Sci 187:85–92. https://doi.org/10.1016/j.applanim.2016.12.004
6. Biswas G, Thirunavukkarasu AR, Sundaray JK, Kailasam M (2010) Optimization of feeding frequency of Asian seabass (*Lates calcarifer*) fry reared in net cages under brackishwater environment. Aquaculture 305:26–31. https://doi.org/10.1016/j.aquaculture.2010.04.002
7. Navarro-Guillén C, Yúfera M, Engrola S (2017) Daily feeding and protein metabolism rhythms in Senegalese sole post-larvae. Biol Open 6:77–82. https://doi.org/10.1242/bio.021642
8. Alanara A (1992) The effect of time-restricted demand feeding on feeding activity growth and feed conversion in rainbow trout (*Oncorhynchus mykiss*). Aquaculture 108:357–368
9. Nakayama S, Johnstone RA, Manica A (2012) Temperament and hunger interact to determine the emergence of leaders in pairs of foraging fish. PLoS ONE 7:e43747. https://doi.org/10.1371/journal.pone.0043747
10. Razman MAM, Susto GA, Cenedese A, Abdul Majeed APP, Musa RM, Abdul Ghani AS, Adnan FA, Ismail KM, Taha Z, Mukai Y (2019) Hunger classification of *Lates calcarifer* by means of an automated feeder and image processing. Comput Electron Agric 163:104883. https://doi.org/10.1016/J.COMPAG.2019.104883
11. Khuller S, Rosenfeld A, Wu A (2000) Centers of sets of pixels. Discret Appl Math 103:297–306. https://doi.org/10.1016/S0166-218X(99)00248-6
12. Cubitt KF, Williams HT, Rowsell D, McFarlane WJ, Gosine RG, Butterworth KG, McKinley RS (2008) Development of an intelligent reasoning system to distinguish hunger states in Rainbow trout (*Oncorhynchus mykiss*). Comput Electron Agric 62:29–34. https://doi.org/10.1016/j.compag.2007.08.010
13. Ariyomo TO, Watt PJ (2015) Effect of hunger level and time of day on boldness and aggression in the zebrafish *Danio rerio*. J Fish Biol 86:1852–1859. https://doi.org/10.1111/jfb.12674
14. Rose CS, Stoner AW, Matteson K (2005) Use of high-frequency imaging sonar to observe fish behaviour near baited fishing gears. Fish Res 76:291–304. https://doi.org/10.1016/j.fishres.2005.07.015
15. Zhao J, Bao WJ, Zhang FD, Ye ZY, Liu Y, Shen MW, Zhu SM (2017) Assessing appetite of the swimming fish based on spontaneous collective behaviors in a recirculating aquaculture system. Aquac Eng 78:196–204. https://doi.org/10.1016/J.AQUAENG.2017.07.008

Chapter 3
Image Processing Features Extraction on Fish Behaviour

Abstract This chapter demonstrates the pipeline from data collection until classifier models that achieve the best possible model in identifying the disparity between hunger states. The pre-processing segment describes the features of the data sets obtained by means of image processing. The method includes the simple moving average (SMA), downsizing factors, dynamic time warping (DTW) and clustering by the k-means method. This is to rationally assign the necessary significant information from the data collected and processed the images captured for demand feeder and fish motion as a synthesis for anticipating the state of fish starvation. The selection of features in this study takes place via the boxplot analysis and the principal component analysis (PCA) on dimensionality reduction. Finally, the validation of the hunger state will be addressed by comparing machine learning (ML) classifiers, namely the discriminant analysis (DA), support vector machine (SVM) and k-nearest neighbour (k-NN). The outcome in this chapter will validate the features from image processing as a tool for identifying the behavioural changes of the fish in school size.

Keywords Dynamic time warping · K-means clustering · Features selection · Classification · Boxplot analysis · PCA

3.1 Overview

The pre-processing data section is used to organize the signal that results in vector form, labelling every meaning filtered via SMA and DTW, examining downsize data points effects and finally validating clustered data before the data set is classified. It would, therefore, provide the following procedures with reasonable and countable data sets. The examination begins by analysing the size of the data collected by the features extracted. Previously, data were collected from the experiment before the classification where the acquisition was established initially at the maximum transmitting speed possible. This provides that throughout data uploading, there will be no losing input.

The filtering of data is done to study the optimum time for gathering information. SMA has been shown to address a simpler method of averaging that could give similar attributes in the aquaculture field [1]. After the average signal resolution, the method

© The Author(s), under exclusive license to Springer Nature Singapore Pte Ltd. 2020
M. A. Mohd Razman et al., *Machine Learning in Aquaculture*,
SpringerBriefs in Applied Sciences and Technology,
https://doi.org/10.1007/978-981-15-2237-6_3

will be analysed in the different sampling rates collected during the image acquisition as the reasoning of an optimal data size could constitute the data on the movements of fish. With the understanding of data sets related to different dimensions, the computational complexity for the subsequent ranking would be further decreased. This is achieved by considering the downsizing variables on which a sample size exploration is dependent. The bonded signals are reduced further by selecting the variable x by keeping the original test and each nth after the first iteration [2].

Additionally, further developments in DTW in the latest years are worth mentioning in which it assesses the resemblance and correcting between two signals [3]. The signal alignment is essential to minimize noise from data sources everyday. Once the conformity of signals has been established, the clustering process takes place as this is to impart the identification of hunger states as this was shown to be dependably credible to justify the responses of the fish [4]. The gathered data sets require further insight in terms of the features selected. Therefore the boxplot analysis and principal component analysis (PCA) will be employed to acquire such knowledge [5, 6]. In subsequent, the validation of these methods will be derived using ML algorithms, namely DA, SVM and k-NN, as these classifiers have shown to recognize the classes predicted for the fish behaviours [7–9]. In conclusion, the methods elaborated would suggest a framework of identifying the efficacy of the fish monitoring system between automated demand feeder and image processing that would produce a comprehensive interpretation of mitigating the ambiguous nature for aquaculture engineering.

3.2   Data Pre-processing

The data set that consists of 850,000 data points is categorized as Day1, Day2 and Day3, respectively. The sampling rate employed for the data collection is 20 Hz. In addition, simple moving average (SMA) was performed on the data set in which subset size, and T is assigned to 20 data points. Moreover, a downsizing of the data set was carried out with a sampling rate of 0.33 and 1 Hz, respectively. In other words, three distinct data sets will be further examined, i.e. the original (20 Hz), 1 and 0.33 Hz. Dynamic time warping (DTW) was carried out to measure the similarity and adjustment made between the two signals. Nonetheless, it is worth noting that there are other advancements made on DTW in recent years [3]. However, in the present investigation, only the original form of DTW is employed, i.e. the warping path of the given boundary conditions is constrained along with $\xi_1 = (1, 1)$ and $\xi_L = (U, V)$. The total warping path could be express as:

$$C_\xi(u, v) = \sum_{l=1}^{L} c(u(i_l), v(j_l)) \tag{3.1}$$

Table 3.1 Steps of k-means

k-means algorithm steps
Initialize: Randomly choose initial centroids $\mu_1, \dots, \mu_k$
Repeat until convergence: $\forall i \in [k]$ set $C_i = \{x \in X : i = \arg\min_j \|x - \mu_j\|\}$ (break ties in some arbitrary manner)
Update: $\forall i \in [k]$ update $\mu_i = \frac{1}{C_i} \sum_{x \in C_i} x$

With the distance for vector, signal between u and v that are associated with ξ is attained from the DTW distance between the components' root square of the sum of squares:

$$D_{\text{DTW}}(u, v) = \min_{\xi}\{C_{\xi}(u, v)\} \qquad (3.2)$$

3.3 Clustering

The k-means algorithm is used to cluster the hunger state of the fish as hungry or not hungry [4]. This distance metrics utilized in this study is the Euclidean distance that could be mathematically expressed as:

$$d(x, y) = \sqrt{\sum_{i=1}^{N}(x_i - y_i)^2} \qquad (3.3)$$

where d is the distances, N is the total number of the data point, and x_i and y_i are the points and mean centroids in that particular cluster, respectively. The steps of the algorithm can be defined as Table 3.1 where the data is partitioned into disjoint sets $C_1, \dots, C_k$ where each C_i is represented by a centroid μ_i. Subsequently, the Silhouette analysis is used to evaluate the efficacy of the cluster size determined by the k-means algorithm [10].

3.4 Feature Selection and Classification

The boxplot analysis and the principal component analysis (PCA) are used for extracting the significant features in the present investigation [11, 12]. This section will provide a brief on the fundamental concepts of both boxplots and PCA on the feature selection basis.

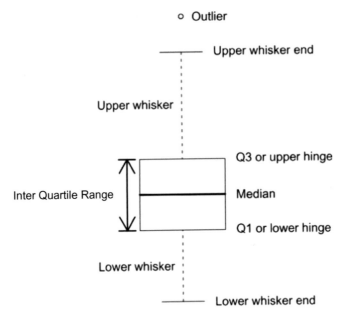

Fig. 3.1 Boxplot annotation. *Source* Seltman [13]

3.4.1 *Boxplot Analysis*

The basic understanding of boxplots is to display the distribution from given data sets in this case from various features that were extracted from the image processing technique. The main criterion of the boxplot is illustrated in Fig. 3.1, in which the five major components are the lower whisker end, lower hinge, median, upper hinge and the upper whisker end [13].

The key indication from the boxplots is the median where the distributions are revolve based on this, and the interquartile range (IQR) which scales between the 25th and 75th for the Q1 and Q3, respectively. On the other hand, lower and upper whisker ends or generally known as the minimum and maximum separate between these two points with the factor of 1.5 times the IQR span. The outlier is customarily the data points that detached with the other distributions or statistically figured at 0.7% of the overall data. From the boxplot analysis, the significant features are identified through the position of the median and the IQR separation.

3.4.2 *Principal Component Analysis (PCA)*

The general objective of PCA is to reduce the dimensionality and focuses on the most significant contribution of the signal [11, 12]. Matrix D as the dimension is to be examined and extracts the orthonormal basis where the variability explained by the original matrix decreases accordingly in the order of the components. The updated elements are the eigenvectors of the matrix D covariances that lead to the first principal components are highly informative. The annotation of the principal component can be described in the following equation:

$$\alpha_1'x = \alpha_{11}x_1 + \alpha_{12}x_2 + \cdots + \alpha_{1p}x_p = \sum_{j=1}^{p}\alpha_{1j}x_j. \qquad (3.4)$$

The equation above expresses as the linear function of $\alpha_1'x$ in which x is the vector for the random variables p that has the highest variance. However, by reducing the dimensionality of the features, it does not interpret the specified features explicitly. Therefore, the varimax rotation is performed to determine the actual features and not transformed features.

3.4.3   Machine Learning Classifiers

Machine learning classifiers are used in the present investigation owing to the erratic nature of the data set, in which conventional form of statistical models could not cater [14]. In this study, discriminant analysis (DA), support vector machine (SVM) and k-nearest neighbour (k-NN) algorithms are investigated towards its efficacy in classifying the hunger state of *Lates calcarifer*.

3.4.3.1   Discriminant Analysis

The DA is assumed to be the conditional probability distribution based on Gaussian function that separates the responses linearly and has been applied in classifying the pigmentation patterns on fish [6]. The type of DA applied in this study is the linear variation. The general concept of DA is to evaluate the competency of given supervised data sets that were allocated in a group of clusters comparatively with others, and it is considered as a generative model [14]. The groups that were formed by the discriminant function can be expressed as Eq. 3.5:

$$f(G_i) = k_i + \sum_{j=1}^{n} w_{ij} P_{ij} \qquad (3.5)$$

where G_i is the group's total number based on the i value, the constant innate for each group is given by k_i, the sum number of variables inserted into specifying each to each group is n and the DA function analysis stipulates the volume coefficient of w_{ij} to the P_j which is the parameter.

3.4.3.2   Support Vector Machine

Conversely, SVM is based on the boundaries defined by the largest distance between classes commonly termed as margin. It is widely applied in different fields either to solve regression or classification problems and was seen to predict high precision in marine biology discipline [15–17]. The earliest practice of SVM was initiated by Cortes and Vapnik in 1995, where they rationalize the fundamental principle of the algorithm theorem [18]. The parameter that is responsible for addressing the classification ability is the objective function that is based on a trade-off between the maximizing and error penalization in the classification [19]. The equation below states the kernel function that employed within a training vector to measure its performance:

$$f(x) = \sum_{i=1}^{l} y_i \alpha_i K(x, x') + b \tag{3.6}$$

where the kernel function of $K(x, x')$ alters the adjustment of the distance to compute the training vector between two classes; besides, the SVM could provide a solution for multi-class classification problems. The algorithm is based on linearized function, but since the hyperparameter tuning optimizer could compute the variation of parameters changes in an algorithm, the radial basis function is also being evaluated as the substitute choice of K denoted below:

$$K(x, x') = \exp(-\gamma ||x - x'||^2) \tag{3.7}$$

where the radial kernel parameter denoted as γ. The hyperparameter will tune the C or the cost function to generate a steady form between the regularization and loss term express in the following equation:

$$\arg \min_{w,\xi} \left\{ \frac{1}{2} ||w||^2 + C \sum_{i=1}^{n} \xi_i \right\} \tag{3.8}$$

where the term of regularization is given by $||w||^2$, and the loss term is the sum of ξ_i.

3.4.3.3   *K-nearest Neighbour*

The k-NN which exploits the distance between the labelled input space of neighbours. The Euclidean distance approach is commonly applied to calculate the distances between each neighbour and was seen to be effective in classifying images of fish to detect the level of freshness [8]. The expression of the distance can be formulated as in Eq. 3.9.

$$\rho(x, x') = ||x - x'|| = \sqrt{\sum_{i=1}^{d}(x_i - x_i')^2} \qquad (3.9)$$

where ρ can be denoted as the Euclidean distance between the point of x_i of new data and x_i' of the training data, and the notation of d is the dimension of the real number set. As for hyperparameter tuning, the number of neighbours, k, and other distance metrics will be evaluated to gauge the minimum cost function of the objective model.

3.5 Results and Discussion

There are several experiments in the context of statistical methods, and machine learning techniques were carried out to evaluate the best methodology to classify the hunger state. The first evaluation is carried out on Experiment A (Exp A) where the sampling size of the data is reduced from 20 Hz (the original sampling time as shown in Fig. 3.2) to 1 Hz and 0.33 Hz, respectively. Reducing the data sets is known to alter the behaviour of the original signals. The objective of this experiment is to determine whether the reduced sample size could still provide a high accuracy prediction. Subsequently, Experiment B (Exp B) is performed by applying SMA and DTW on the reduced data set.

It is hypothesized that the k-means clustering technique would cluster the state of hunger as either 'Hungry' or 'Not Hungry'. DTW is then implemented for synchronizing the phases between separate data sets used, in this case, different days of data that was acquired. Figure 3.2 demonstrates the aligned data sets between Day1 and Day2. Table 3.2 provides the performance of the k-means algorithm by varying the

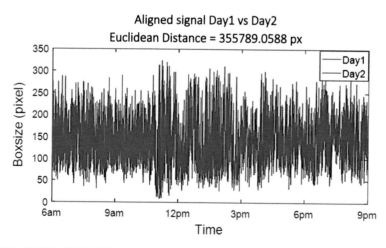

Fig. 3.2 DTW for BOX_SIZE feature between the first day and second day

Table 3.2 Comparing *k*-means for different sampling time on the first-day data set

k-means clusters	Data type		
	20 Hz	1 Hz	0.33 Hz
	Norm	SMA	SMA-DTW
k = 2	0.7374	0.7386	0.4996
k = 3	0.6774	0.6798	0.4306
k = 4	0.6493	0.6493	0.4304

number of *k*, while using the Euclidean distance on the number of optimum classes, it could discriminate based on the BOX_SIZE feature via the Silhouette score on the variation of data set prepared. It is evident from the table that regardless of the type of data set, two clusters (*k* = 2) appear to be dominant across the data set, suggesting that the hunger state indeed could be demarcated as either 'Hungry' or 'Not Hungry'.

Therefore, the two clusters, i.e. 'Hungry' and 'Not Hungry' (satiated), will be applied for evaluating the aforesaid classifier models in the present study. However, the determination of the hunger state and satiated states will only be ascertained through the boxplot analysis that will be subsequently deliberated. From these results, classification methods on the supervised learning models will be adopted.

Figure 3.3 displays the classification accuracy between both 'raw' that invariably applied SMA and the DTW (customarily applying SMA as well) test data sets for Day2. All the features described in Chap. 2 are used for this analysis. It could be seen from the figure that the DTW data set could provide a better classification accuracy in comparison with the 'raw' data set. Among the classifiers investigated, it is evident that the SVM classifier provided the best classification accuracy that is up to 97.9%.

Figure 3.4 illustrates the rescaled boxplots of various features evaluated and their deviation between classes. Previous preliminary study has established that the lower

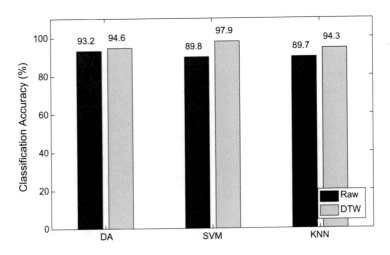

Fig. 3.3 Classification accuracy of test data the second day

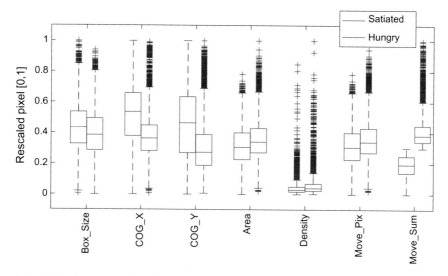

Fig. 3.4 Boxplots comparisons in a scale [0 1] for all the features

the *COG_Y* indicates that the school of fish is satiated [20], and this is evidently demonstrated from Fig. 3.4. In addition, it is also apparent that *COG_X*, as well as the *Move_Sum features,* are significant as also remarked in [21, 22]. Figure 3.5 illustrates the three selected features. Herein, the term 'normalize' is denoted as

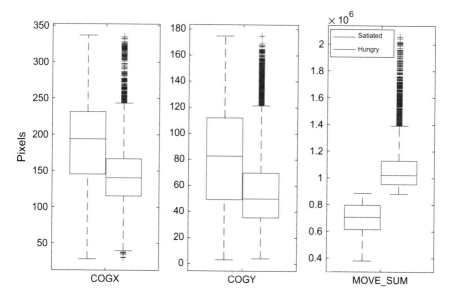

Fig. 3.5 Pixels scale for selected features

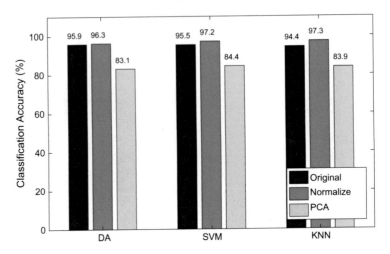

Fig. 3.6 Test data classification accuracy of original, normalize and PCA

the aforesaid selected three features. Subsequently, the effect of PCA, normalize as well as the DTW version of the data set which is denoted as 'original' towards the classification accuracy of different classifiers are investigated. It could be seen from Fig. 3.6 that the 'normalize' data provides a better accuracy as compared to the 'original' and the PCA set for all classifiers. It is also established from the present investigation based on all the 'normalize' data set, and the k-NN classifier attained the best classification accuracy among the models evaluated.

3.6 Summary

This chapter has demonstrated a pipeline for the hunger classification of *Lates calcarifer*. It was shown that the effect of downsampling of the data reduced the data set significantly, and from the analyses carried out, the DTW method could provide a good classification accuracy. In addition, the significant features are extracted by means of boxplot analysis as well as PCA. It was shown from the boxplot analysis that *COG_X*, *COG_Y* and *Move_Sum* could provide a reasonable classification of hunger state, and the SVM model could discriminate the hunger state better than the other models evaluated.

References

1. Scandol J (2005) Use of quality control methods to monitor the status of fish stocks. In: G K (ed) Fisheries assessment and management in data-limited situations. Alaska Sea Grant, University of Alaska Fairbanks, pp 216–266
2. Guan J, Liu H, Shi X, Feng S, Huang B (2017) Tracking multiple genomic elements using correlative CRISPR imaging and sequential DNA Fish. Biophys J 112:1077–1084. https://doi.org/10.1016/J.BPJ.2017.01.032
3. Morel M, Achard C, Kulpa R, Dubuisson S (2018) Time-series averaging using constrained dynamic time warping with tolerance. Pattern Recogn 74:77–89. https://doi.org/10.1016/J.PATCOG.2017.08.015
4. Yassin W, Rahayu S, Abdollah F, Zin H (2016) An improved malicious behaviour detection via k- means and decision tree. IJACSA Int J Adv Comput Sci Appl 7
5. Kennedy J, Jónsson SÞ, Ólafsson HG, Kasper JM (2016) Observations of vertical movements and depth distribution of migrating female lumpfish (*Cyclopterus lumpus*) in Iceland from data storage tags and trawl surveys. ICES J Mar Sci J du Cons 73:1160–1169. https://doi.org/10.1093/icesjms/fsv244
6. Wishkerman A, Boglino A, Darias MJ, Andree KB, Estévez A, Gisbert E (2016) Image analysis-based classification of pigmentation patterns in fish: a case study of pseudo-albinism in *Senegalese sole*. Aquaculture 464:303–308. https://doi.org/10.1016/J.AQUACULTURE.2016.06.040
7. Hasija S, Buragohain MJ, Indu S (2017) Fish species classification using graph embedding discriminant analysis. In: 2017 international conference on machine vision and information technology (CMVIT). IEEE, pp 81–86
8. Iswari NMS, Wella, Ranny (2017) Fish freshness classification method based on fish image using k-Nearest Neighbor. In: 2017 4th international conference on new media studies (CONMEDIA). IEEE, pp 87–91
9. Razman MAM, Susto GA, Cenedese A, Abdul Majeed APP, Musa RM, Abdul Ghani AS, Adnan FA, Ismail KM, Taha Z, Mukai Y (2019) Hunger classification of *Lates calcarifer* by means of an automated feeder and image processing. Comput Electron Agric 163. https://doi.org/10.1016/j.compag.2019.104883
10. Gastauer S, Scoulding B, Parsons · Miles An Unsupervised Acoustic Description of Fish Schools and the Seabed in Three Fishing Regions Within the Northern Demersal Scalefish Fishery (NDSF, Western Australia). Acoust Aust. https://doi.org/10.1007/s40857-017-0100-0
11. Broell F, Noda T, Wright S, Domenici P, Steffensen JF, Auclair J-P, Taggart CT (2013) Accelerometer tags: detecting and identifying activities in fish and the effect of sampling frequency. J Exp Biol 216:1255–1264. https://doi.org/10.1242/jeb.077396
12. Valletta JJ, Torney C, Kings M, Thornton A, Madden J (2017) Applications of machine learning in animal behaviour studies. Anim Behav 124:203–220. https://doi.org/10.1016/J.ANBEHAV.2016.12.005
13. Seltman HJ (2018) Experimental Design and Analysis. Carnegie Mellon University
14. Shalev-Shwartz S, Ben-David S (2013) Understanding machine learning: From theory to algorithms
15. Dutta MK, Sengar N, Kamble N, Banerjee K, Minhas N, Sarkar B (2016) Image processing based technique for classification of fish quality after cypermethrine exposure. LWT - Food Sci Technol 68:408–417. https://doi.org/10.1016/j.lwt.2015.11.059
16. Cubitt KF, Williams HT, Rowsell D, McFarlane WJ, Gosine RG, Butterworth KG, McKinley RS (2008) Development of an intelligent reasoning system to distinguish hunger states in Rainbow trout (Oncorhynchus mykiss). Comput Electron Agric 62:29–34. https://doi.org/10.1016/j.compag.2007.08.010
17. Ogunlana SO, Olabode O, Oluwadare SAA, Iwasokun GB (2015) Fish classification using support vector machine. Afr J Comput ICT Afr J Comput ICT Ref Format Afr J Comp ICTs 8:75–82

18. Cortes C, Vapnik V (1995) Support-vector networks. Mach Learn 20:273–297
19. Chang C-C, Lin C-J (2011) LIBSVM. ACM Trans Intell Syst Technol 2:1–27. https://doi.org/
    10.1145/1961189.1961199
20. Taha Z, Razman MAM, Adnan FA, Abdul Majeed APP, Musa RM, Abdul Ghani AS, Sallehudin
    MF, Mukai Y (2018) The identification of hunger behaviour of *Lates calcarifer* using k-nearest
    neighbour. Springer, Singapore, pp 393–399
21. Cavallari N, Frigato E, Vallone D, Fröhlich N, Lopez-Olmeda JF, Foà A, Berti R, Sánchez-
    Vázquez FJ, Bertolucci C, Foulkes NS (2011) A blind circadian clock in cavefish reveals that
    opsins mediate peripheral clock photoreception. PLoS Biol 9:e1001142
22. Alós J, Martorell-Barceló M, Campos-Candela A (2017) Repeatability of circadian behavioural
    variation revealed in free-ranging marine fish. R Soc Open Sci 4:160791. https://doi.org/10.
    1098/rsos.160791

Chapter 4
Time-Series Identification on Fish Feeding Behaviour

Abstract The identification of relevant parameters that could describe the state of fish hunger is vital for ensuring the appropriate allocation of food to the fish. The establishment of these relevant parameters is non-trivial, particularly when developing an automated demand feeder system. The present inquiry is being undertaken to determine the hunger state of *Lates calcarifer*. For data collection, a video analysis system is used, and the video was taken all day, where the fish was fed by an automatic feeding system. Sixteen characteristics of the raw data set have been extracted through feature engineering for 0.5 min, 1.0 min, 1.5 min and 2.0 min, respectively, in accordance with the mean, peak, minimum and variability of each of the different time window scales. Furthermore, the features extracted have been evaluated through principal component analysis (PCA) both for dimension reduction and PCA with varimax rotation. The details were then categorized using support vector machine (SVM), K-NN and random forest tree (RF) classifiers. The best identification accuracy was shown with eight described features in the varimax-based PCA. The forecast results based on the K-NN model built on selected data characteristics showed a level of 96.5% indicating that the characteristics analysed were crucial to classifying the actions of hunger among fisheries.

Keywords Image processing · Automated demand feeder · *Lates calcarifer* · Pixel intensity · Specific growth rate

4.1 Overview

Primarily, a group of fish exhibits higher movement when searching for food which often describes their state of hunger while the movement tends to decline or reduces when the fish is satiated [1, 2]. Therefore, the swimming ability of the fish may fluctuate with regard to the hunger level of the fish. A research has shown that fish appear to be more aggressive and cover the field when they are starving and therefore have a greater movement as individuals and in a shoal [3]. Another experiment has shown that hunger can be caused by access to fish to the ultradian pattern of light and darkness replicating pulses every day and night [4]. This scenario implies that the formation of the time interval has to be taken into account. However, it is challenging

© The Author(s), under exclusive license to Springer Nature Singapore Pte Ltd. 2020
M. A. Mohd Razman et al., *Machine Learning in Aquaculture*,
SpringerBriefs in Applied Sciences and Technology,
https://doi.org/10.1007/978-981-15-2237-6_4

to state the accurate period as the moment of time and day could alter the fish hunger behaviour couple with other endogenous as well as exogenous influences. Therefore, a time-series analysis has to be employed to mitigate the aforesaid issue which could be addressed by means of feature engineering extraction technique obtain through image processing.

The present chapter is structured in sequential order. The first stage is aimed at identifying the occurrence of events involved in the state of fish hunger behaviour between two classes as examined earlier. The second stage comprised of the dissolution of the previously classified group in order to gauge the appropriate time window for the purpose of addressing the second layer features. The PCA method is then implemented to provide insights on the optimal features that could be used in describing the hunger behaviour of the fish. Subsequently, the SVM model is developed for validating the selected features while hyperparameter tuning is carried out to optimize the classification accuracy of the model. Moreover, correlation matrices, as well as scree plots, were applied to identify the significant features that contribute towards the formation of the effective ML models that best explain the hunger behaviour of the fish in the present investigation. It is worth highlighting that the optimization analysis section is essentially formulated to aid in the identification of the best parameters that could be used to improving both the accuracy as well as the run-time of the model developed in the present investigation.

4.2 Event Identification

It is essential to choose the instances involved in the behavioural changes of fish hunger in order to permit the labelling of the responses exhibited according to the classes that were previously clustered. The often produced performance from the cluster analysis is only capable of producing the best number of clusters as regards data sets; as such the marking of specific and real groups needs to be done on the basis of an expert in the field of fish behaviour. The details in the time series obtained in this analysis are therefore in a real-world situation in which the parameters differ with the period stated below:

$$y(t), \ t = 1, 2, \ldots, n \tag{4.1}$$

where the parameters or features extracted are $y(t)$, and time variable represented by t [5]. The known changes in this study lie when the fish is fed for a given time and as it reaches full satiated phase the behaviour changes to a stagnant motion or as portray by the automated demand feeder is when the activation from the trigger sensor is stopped. One can then conclude that the responses can now be treated as 'Satiated' and can be classed as 'Hungry' when the sensor is activated.

4.3 Features Selection PCA-Based

As a time sequence issue is understood, it can be represented when features in second layers if behavioural variations are accomplished. The answers can be identified by analysing the activity period or the circadian rhythm moment of famine [6]. P feature extraction technique is employed to extract the window size of the features. In this step, the implication of applying p features is demonstrated. The input is expressed from $x \in \mathbb{R}^{1 \times p}$ is gained by the several features in time series as below:

$$a(t) = \mathbb{R}^{n \times p}, \ t = 1, \ldots, U \tag{4.2}$$

where $\mathbb{R}$ is the actual number collection, n is the number of sizes selected and p is the number of features of the second level derived from video processing capability. For example, a 0.5 min window has four features (mean, max, min, var) which show the first p functionality with four features inside a p. Through adding a further 1.0 min window, the p features are raised to $2p$. Hence, 0.5, 1.0, 1.5 and 2.0 min window dimensions are collected in this case, which indicates that sample size should add $4p$ characteristics, as shown in this case $x \in \mathbb{R}^{m \times 4p}$. In addition, it results in 16 features in full. The window size is where the detection of movement is being generalized in which the speed of hunger or any sudden changes or abruption could be detected, the steadiness of the group and the outliers of individual fish from the group could, therefore, be determined.

As far as the PCA concept is concerned, the varimax rotation involves recognition of variables dependent on the principal components' own values greater than 1. For each element, the associated function is the varimax factor, which defines the unknown, conjectural and indistinct variables. The variables are positive or negative and find all levels hitting up to 1 and -1 to be a high correlation [7].

The labelling of the actual classes of the fish in relation to the hunger behaviour could be determined from the motion of the fish by means of event identification technique. In the time-series analysis, the second layer features analysis could offer information on the circadian rhythm of the fish hunger behaviour. At this juncture, the technique of classifying the hunger behaviour has been exemplified for identifying the fish behaviour through image processing features, translating the features, analysing the significant parameters as well as verifying the period instances of the movement. The processed data is then classified through the application of a variety of ML models, and comparative analysis is conducted to ascertain the best predictive model of the fish hunger behaviour in the present study.

4.4 Classification Accuracy

In this research SVM, k-NN and RF are tested as identification templates. This paper does not expand on the specifics of these designs. The readers are nevertheless urged to consider the previous literature [8]. SWM builds on the limits of categories commonly referred to as ranges as the largest size. The variable of classification is a feature based on a balance between the optimizing and the penalizing mistake in the classification equation [9].

On the other side, the k-NN method utilizes the range among neighbour's marked input spaces. The method to distance between Euclidean's is often used to measure distances between adjacent individuals and was shown to be successful in classifying fish pictures for freshness detection [10].

The RF system has been developed to improve the accuracy level of classification for traditional decision trees often called classification and regression trees (CART) by sharing the node or by arbitrarily searching for the responses [11]. This design is preferable to the previous version because it reduces the overfitting possibilities. The concept of assembling the trees that satisfy enough for the forest; in other words, the entire model will reduce the overfitting of the classifier. Furthermore, it has the advantages of managing the uncertainties value that is lost during searching, and the model could be described in a definite model.

In this analysis, box constraint, c and kernel scale, hyperparameters are chosen as 1 for the linear SVM, while for k-NN the Euclidean distance with the number of neighbours chosen as 1 is used for the Euclidean size.

The criterion of classification accuracy with the percentage of the predicted responses has been correctly classified by computing the confusion matrix that points out the misclassification rate distributed onto the respective classes [12]. The variants of the ML models used in this analysis were tested using reliability of identification, recall or sensitivity and precision. Figure 4.1 demonstrates the confusion matrix as an evaluation of the predictive models.

As illustrated in Fig. 4.1, the confusion matrix evaluates the respective results from the actual against the predicted outcome in terms of accuracy, precision, recall and F_1 score. The understanding of the matrix outcome would be better explained in the analogy of this study to classify hunger or satiated state. For instance, the True Positive (TP) stands when predicting the fish as hungry; however, the False

Fig. 4.1 Confusion matrix table

Confusion Matrix		Predicted		
		True	Hungry	Satiated
Actual	Hungry		True Positive	False Negative
	Satiated		False Positive	True Negative

Positive (FP) when the hunger state has been labelled into the satiated phase. These conditions conversely followed by the False Negative (FN) in which the expected state of the satiation behaviour is incorrectly labelled as hungry. Nonetheless, the True Negative (TN) accurately marks the satiated responses. The instances generated by the confusion matrix for the accuracy, precision, recall and F_1 score can be expressed through the following equations:

$$Accuracy = \frac{TP + TN}{TP + TN + FP + FN} \tag{4.2}$$

$$Precision = \frac{TP}{TP + FP} \tag{4.3}$$

$$Recall = \frac{TP}{TP + FN} \tag{4.4}$$

$$F_1 = 2 \cdot \frac{Precision \cdot Recall}{Precision + Recall} \tag{4.5}$$

From the stated equations, the accuracy stands as the overall classification rate on the regularity of the classifier in predicting the classes, precision evaluates the prediction rate of being recurrent yes or correctly predicted. The recall is the fractions of yes being labelled or the positive that is correctly categorized. The F_1 is the harmonic mean that is generated between precision and recall by multiplying the scale by 2. The F_1 score suggests the low volume of classes that are being falsely labelled and hence shows the strength of the prediction from the overall accuracy.

4.5 Results and Discussion

A total number of 59,807 were acquired throughout four days experiment. Both the class identification and the feeding activity dependent on COGy and box size are assessed. The option of these two characteristics among the seven is focused on the recommendation of previous studies [8, 13]. The general intention of the analysis is to measure the hunger between the satiated and the deprived, often the target of a mere human decision that can be skewed and misunderstood. The process begins by removing the fish feed time from the automated feeder until it ceases requesting. Here it is believe that it can be calculated by identifying the feeding cycle that is 'Hungry' or 'Satiated'. For example, since the feed period was at 11:00 am, the feed ends at about five minutes as the rate of COGy is declining, while the pits rise at 11:10 am where the feed time is shown in Fig. 4.2. It indicates that the behaviour changes from hunger to satiation. The feeding cycle of the data set was not included, and only the starving and satiated condition was left. It means that both groups are collected for the corresponding controlled sample evaluation [14].

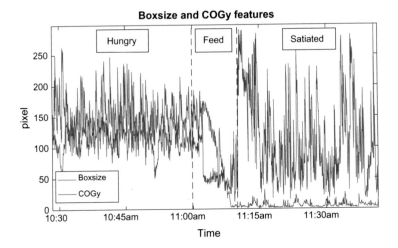

Fig. 4.2 Temporal window extraction for feeding process

The scree interpretation of the PCA study presented in Fig. 4.3 shows that the results that the first two components add up a high percentage of variance in which the quality of the particular principal is higher than one and the corresponding variable axis is suggested [12]. The first variable charging shows the value of 11.2 with the percentage variance of about 73%, and then the second with 3.1 eigenvalue, and thus increasing the percentage of total volatility to approximately 90%. The outcomes

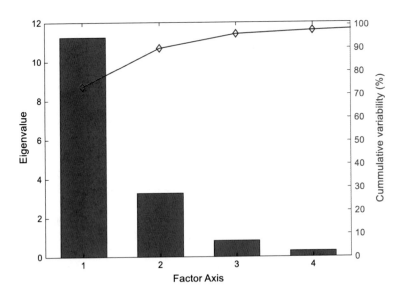

Fig. 4.3 Cumulative variability and eigenvalues scree plot

Table 4.1 Factor loading after varimax rotation

Features		Components	
		C1	C2
0.5 min	Mean	0.97	−0.04
	Max	0.65	−0.64
	Min	0.65	−0.64
	Var	0.07	−0.88
1.0 min	Mean	0.98	−0.04
	Max	0.70	−0.68
	Min	0.70	−0.68
	Var	0.07	−0.96
1.5 min	Mean	0.98	−0.04
	Max	0.71	−0.66
	Min	0.71	−0.66
	Var	0.06	−0.96
2.0 min	Mean	0.98	−0.03
	Max	0.71	−0.62
	Min	0.71	−0.62
	Var	0.05	−0.92

show that both major components are equally important when defining key features in the data set.

Table 4.1 points out the chosen element features from C1 and C2 and illustrates the results of factor loadings on the characteristics tested during varimax rotation. In defining hunger behaviour, bolded typeface values are considered significant. It is clear because, as indicated by the literature, the mean and variance for all windows returned important characteristics [15].

The PCA study concluded that a maximum of eight characteristics are necessary for assessing the fish's hunger behaviour from the initial data developed, while the PCA-based varimax rotation found that two dimensions were sufficient to describe about 90% of the whole data set variability. Hence, the mean square error (MSE) was calculated by SVM classifier to calculate the discrepancy between PCA, PCA with varimax rotation and the 16 features mentioned in Table 4.2 for each function variance.

The MSE percentage and the uncertainty matrix of the three different features that the SVM template assesses are provided in Table 4.2. The eight features from the PCA-based varimax rotation could be observed to show the smallest MSE at 17.40%. It is important to note that the MSE of the 16 features varies little from the two-dimension obtained from the generic PCA study. As a consideration, further analysis of the new data is needed to establish the efficacy for the chosen features of the SVM system.

Table 4.2 SVM classifier accuracy on train data set

(MSE) %		Train data set					
		Two-dimension (2D)		Eight features (Varimax)		16 features (All)	
Error		17.42% ± 0.0003%		17.40% ± 0.0005%		17.79% ± 0.0008%	
Confusion matrix (%)	True	H	S	H	S	H	S
	H	0.77	0.23	0.76	0.24	0.77	0.23
	S	0.12	0.88	0.12	0.88	0.13	0.87

Table 4.3 SVM classifier accuracy on test data set

(MSE) %		Test data set					
		Two-dimension (2D)		Eight features (Varimax)		16 features (All)	
Error		17.53%		17.36%		18.34%	
Confusion matrix (%)	True	H	S	H	S	H	S
	H	0.77	0.23	0.77	0.23	0.76	0.24
	S	0.14	0.86	0.13	0.87	0.14	0.86

Table 4.3 demonstrates the comparative reliability between variants relying on chosen characteristics, i.e. eight and seventeen. With eight features, it is clear that the deviation is decreased from 17.4% of training predictions to 17.36% for test error. The evidence presented here indicates that the template of eight characteristics best forecasts on a fresh data set while the 16 characteristics suggest MSE development from 17.79 to 18.34%. Such results combined support a clearer overview of the eight features than the 16 features for SVM models.

Figure 4.4 shows the variations of the data sets among all classifiers involved. Implementing the single PCA to minimize the dimensionality of the data set does not significantly affect the reliability of classification [16]. For RF and k-NN classifiers, it shows that the varimax set has lower marginal error with 7.6% and 3.5%, respectively, as opposed to two-dimensional PCA with 17.0% and 20.5%, respectively. Nevertheless, it is clear from this work that the use of PCA with varimax rotation factor loads contributed to a higher identification performance than the traditional PCA methodology. Even though the classification reliability of PCA varimax and the choice of all features are shown to be poor, it is important to note that around half of the information is decreased and that the processing time used for non-trivial real-time classification decreases dramatically [17–19]. The k-NN method gives, ultimately, 3.5% error of the system tested by choosing features using PCA with a varimax rotation, the highest classification level.

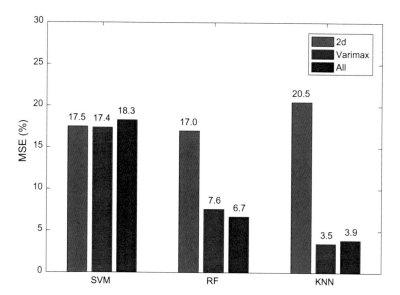

Fig. 4.4 Classification accuracy on the test set

4.6 Summary

Each section has developed a number of applicable features from the different window sizes in which 16 features have been extracted. The PCA evaluation was performed to establish the main characteristics which could characterize fish starvation. In order to extract different characteristics depending on a variable load level, PCA with varimax rotation was also used. In order to evaluate the identification effectiveness of hunger behaviour in relation to above-described functional variance SVM, k-NN, RF models are established. Through the presenting study, it was observed that the PCA-based varimax rotation by k-NN classifier with eight features could well define hunger in comparison with the different models. The approach suggested is necessary in the knowledge of the circadian rhythm of the *Lates calcarifer* coupled with the feeding schedule, which therefore would not only assist in the laboratory but also in the development and production of highly demanded species [20]. It is important to stress that the system built can be studied beyond and contrasted with other recent models including deep learning, augmented learning and the generative opponent network among many others. For real-time automatic fish feeding systems, it can also be transferred to a microcontroller.

References

1. Kennedy J, Jónsson SÞ, Ólafsson HG, Kasper JM (2016) Observations of vertical movements and depth distribution of migrating female lumpfish (*Cyclopterus lumpus*) in Iceland from data storage tags and trawl surveys. ICES J Mar Sci J du Cons 73:1160–1169. https://doi.org/10.1093/icesjms/fsv244

2. Nakayama S, Johnstone RA, Manica A (2012) Temperament and hunger interact to determine the emergence of leaders in pairs of foraging fish. PLoS ONE 7:e43747. https://doi.org/10.1371/journal.pone.0043747

3. Chapman BB, Morrell LJ, Krause J (2010) Unpredictability in food supply during early life influences boldness in fish. Behav Ecol 21:501–506

4. Sanchez-Vázquez FJ, Madrid JA, Zamora S (1995) Circadian rhythms of feeding activity in sea bass, *Dicentrarchus labrax* L.: dual phasing capacity of diel demand-feeding pattern. J Biol Rhythms 10:256–266. https://doi.org/10.1177/074873049501000308

5. Guralnik V, Srivastava J (1999) Event detection from time series data. In: Proceedings of the fifth ACM SIGKDD international conference on Knowledge discovery and data mining—KDD'99. ACM Press, New York, New York, USA, pp 33–42

6. Dutta MK, Sengar N, Kamble N, Banerjee K, Minhas N, Sarkar B (2016) Image processing based technique for classification of fish quality after cypermethrine exposure. LWT Food Sci Technol 68:408–417. https://doi.org/10.1016/J.LWT.2015.11.059

7. Jolliffe IT (2002) Principal component analysis, 2nd edn. Springer Ser Stat vol 98, p 487. 10.2307/1270093

8. Razman MAM, Susto GA, Cenedese A, Abdul Majeed APP, Musa RM, Abdul Ghani AS, Adnan FA, Ismail KM, Taha Z, Mukai Y (2019) Hunger classification of *Lates calcarifer* by means of an automated feeder and image processing. Comput Electron Agric 163:104883. https://doi.org/10.1016/J.COMPAG.2019.104883

9. Chang C-C, Lin C-J (2011) LIBSVM. ACM Trans Intell Syst Technol 2:1–27. https://doi.org/10.1145/1961189.1961199

10. Iswari NMS, Wella, Ranny (2017) Fish freshness classification method based on fish image using k-Nearest Neighbor. In: 2017 4th international conference on new media studies (CONMEDIA): 8–10 Nov 2017, Yogyakarta, Indonesia. IEEE, pp 87–91

11. Breiman L (2001) Random forests. Mach Learn 45:5–32. https://doi.org/10.1023/A:1010933404324

12. Dangeti P (2017) Statistics for machine learning build supervised, unsupervised, and reinforcement learning models using both Python and R. Packt Publishing

13. Mukai Y, Tan NH, Khairulanwar M, Chung R, Liau F (2016) Demand feeding system using an infrared light sensor for brown-marbled grouper juveniles, *Epinephelus fuscoguttatus*. Sains Malaysiana 45:729–733

14. Broell F, Noda T, Wright S, Domenici P, Steffensen JF, Auclair J-P, Taggart CT (2013) Accelerometer tags: detecting and identifying activities in fish and the effect of sampling frequency. J Exp Biol 216:1255–1264. https://doi.org/10.1242/jeb.077396

15. Peré-Trepat E, Olivella L, Ginebreda A, Caixach J, Tauler R (2006) Chemometrics modelling of organic contaminants in fish and sediment river samples. Sci Total Environ 371:223–237. https://doi.org/10.1016/j.scitotenv.2006.04.005

16. Tian XY, Cai Q, Zhang YM (2012) Rapid classification of hairtail fish and pork freshness using an electronic nose based on the PCA method. Sensors 12:260–277. https://doi.org/10.3390/s120100260

17. Vanaja S, Ramesh Kumar K (2014) Analysis of feature selection algorithms on classification: a survey. Int J Comput Appl 975–8887

18. Hira ZM, Gillies DF (2015) A review of feature selection and feature extraction methods applied on microarray data. Adv Bioinf. https://doi.org/10.1155/2015/198363

19. Bania RK (2014) Survey on feature selection for data reduction. Int J Comput Appl 94:975–8887

20. Zhao J, Bao WJ, Zhang FD, Ye ZY, Liu Y, Shen MW, Zhu SM (2017) Assessing appetite of
    the swimming fish based on spontaneous collective behaviors in a recirculating aquaculture
    system. Aquac Eng 78:196–204. https://doi.org/10.1016/J.AQUAENG.2017.07.008

Chapter 5
Hyperparameter Tuning of the Model for Hunger State Classification

Abstract To increase the classification, the rate of prediction based on existing models requires additional technique or in this case optimizing the model. Hyperparameter tuning is an optimization technique that evaluates and adjusts the free parameters that define the behaviour of classifiers. Data sets were classified practical with classifiers like SVM, k-NN, ANN and DA. To further improve the design efficiency, the secondary optimization level called hyperparameter tuning will be further investigated. DA, SVM, k-NN, decision tree (Tree), logistic regression (LR), random forest tree (RF) and neural network (NN) are evaluated. The k-NN provided 96.47% of the test sets with the best reliability in classifications. Bayesian optimization has been used to refine the hyperparameter; hence, standardize Euclidean distance metric with a k value of one is the ideal hyperparameters which could achieve classification performance of 97.16%.

Keywords K-nearest neighbour · Neural network · Hyperparameter tuning · Bayesian optimization · Classification

5.1 Overview

The modulation of hyperparameters is an optimization technique in order to determine and adjust the free parameters defining the behaviour of classifier [1]. The hyperparameter optimization techniques can be seen as second-level optimizers as they further improve the accuracy of the classifier which overlaps the key training algorithm. Nevertheless, the sequence of model-based optimization (SMBO) was set for free variables, but the quest for the analysis of the objective variable was a solution to the problem [2].

5.2 Optimization of Bayesian

A Gaussian process design, maintained by Bayesian optimization, determines the minimization of the objective variable. The hyperparameter formula is usually shown as follows:

$$x^* = \arg\min_{x} f(x) \qquad (5.1)$$

If Bayesian optimization attempts to locate the input x^*, which minimizes the feature of $f(x)$ by a certain number of iterations, the estimation of hyperparameter computational measures should initially start with the entry and validation of the data sets by means of the learning and losing algorithm. The iteration then adapts to the Gaussian method, whereby it maximizes to the next hyperparameters through the mean and variance functions. The system is conditioned, and the validity is determined to obtain the loss function after choosing the hyperparameters. Observations are then modified as the template is still given through iterations. Compared with the untune version, the better design from the classification tools will be measured as a test of progress in the assessment of hunger for the school trout. The precision percentages will be deliberated as a requirement to achieve the previously calculated identification level.

As for the previous chapter, the design should contrast eight PCA features with varimax rotation and sixteen features with other settings. The data sets are split into three train sections, with the ratio of 70:15:15 respectively being checked and validated. The design of the models starts by increasing the target variable and by testing the template function. The features are then analysed to determine the objective function approximation for the evaluation of the produced variables. The Bayesian optimization techniques will be used for the hyperparameter tuning in tandem with the iterative solver SMO as the quest framework for the least objective variable. The run-time is also being examined to calculate the classification as an indicator of the model. The best possible template found is mentioned in the last paragraph and listed on the test criteria. The progress in the application of hyperparameter tuning will be seen in this section as an optimiser for classification models.

5.3 Classification Accuracy

For the accuracy calculation, correlations between ML classifiers are evaluated using Bayesian using the best model to optimize. In contrast to the four earlier models DA, SVM, k-NN and RF, seven different models were related with decision tree (Tree), logistic regression (LR) and neural network (NN). This system is assumed to be an ML classifier since, for example, the classes involved are determined by roaming from a tree root node down to other leaves until the target value is achieved [3]. The classifications and regression trees (CART) approach can often be compared as the

Fig. 5.1 Decision tree model assumption on fish hunger

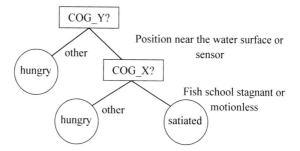

system where descriptors are linked to the categories. The description of the lower class search in accordance with this review is shown in Fig. 5.1. The illustration in the figure shows the descriptors obtained in the previous section of this chapter from image processing techniques. That root, or branch, determines a hungry or satiated answer to the location of the fish school that is defined by the COG Y. The next node is to search the COG X if the motion of the fish has not abstained and satiated if it is on the bottom of the water. The CART system has been shown to be effective for the identification and evaluation of fish using a hydroacoustic method [4] (Fig. 5.1).

LR ignores the context of the distribution of the factors [5]. The LR concept, extend the relationship between the answer and the parameters of the feature, is the most suitable template. In the estimation of the occurrence of fish dependent on locale data, the resiliency of the LR implementation has been shown to be effective [6]. The LR model definition is that the hypothesis of the sets as is a probability of 1 which compose to a sigmoid function over a linear function. Specifically, The LR classifier term that comprises a sigmoid feature is described in general as the logistical function mentioned in Eq. (5.2).

$$\phi_{sig}(z) = \frac{1}{1 + \exp(-z)} \tag{5.2}$$

where in the $\phi_{sig}(z)$ is the sigmoid function that rangers from 0 to 1 as an estimation of the probability of the outcome and z are the input parameters to the function. The LR model's discretization system utilizes a criterion for deviating responses that render distinguishing between groups. The decision boundary has the ability to forecast this outcome appropriately since an illustration could be given as follows, where the groups are between the hungry and satiated:

$$p \geq 0.5, \ \ hungry = 0$$
$$p < 0.5, \ \ satiated = 1 \tag{5.3}$$

The likelihood association between two groups is extended as an above formula would, for example, the template returns a value of 0.3. Thus, the LR method can be evaluated rationally on these research data sets, since only two reactions can be classified according to the original hypothesis, whether starving or satiating.

Table 5.1 Parameters of the classification models

Classifier	Design parameter
Tree	Max splits $= 100$
DA	Type (linear)
LR	$\lambda = 1$
SVM	$C = 1, \gamma = 1$
k-NN	Neighbour $= 1$, distance $=$ euclidean
RF	Num. of trees $= 50$
NN	10 nodes

The NN system was focused on a linked human brain network designed to identify associations in data sets [7]. This comprises of several levels resulting from the predicted outcome and responses of several nodes intertwined with corresponding layers. Adaptation to NN is generally associated with the question of identification as it has seen a development towards fundamental learning difficulties, in general with regard to the organization of imagery [8]. The integration of the layers between input, hidden, and output works in the form of a chain by which the original features or inputs are amplified by the weights frequently found in the hidden layer. The input parameters are fed into the activation function before a meaningful output could be produced. By combining the sigmoid function layers throughout the hidden layer and the softmax function in the outer layer, the classification of hunger behaviour throughout this study is adequate. In essence, the softmax function acquires an unnormalized matrix from data sets and converts it into a probability distribution. Accuracy assessment is expressed in the proportions of the level of identification between the planned and real responses.

Classifiers that have so far been identified have been strictly reflective on the fundamental nature of their respective models. To justify the behaviour classification, a software training and testing system are needed that compares the cross validity between the instances before being evaluated by a misunderstanding matrix. Initially, the parameters of each system are configured, as shown in Table 5.1.

The table displays the classifiers used alongside with the design parameters chosen to differentiate the reliability of the system. All of the models were held at their default state, e.g., the LR model was set at $\lambda = 1$ notation likewise for the SVM system, which was configured with box limit, C, and the kernel size, γ, with a value of one. Likewise, the k-NN system extended only to one set of neighbours and the distance parameter implemented was the Euclidean distance.

5.4 Results and Discussion

Figure 5.2 indicates the correlation of the precision of each classifier with the parameters that have been tabulated, as shown in Table 5.1, where the dominant variables are illuminated. The formats used are the eight characteristics of PCA and varimax

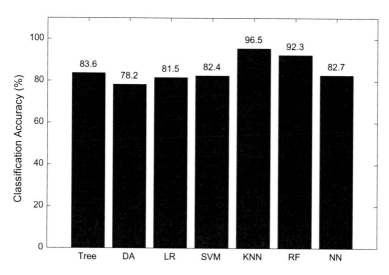

Fig. 5.2 Classification accuracy for various models on the test set

rotation. The variations in percentages shown in Fig. 5.2 indicate that the DA template has the worst accuracy rate of 78.2%. The SVM model has a classification precision of 82.4%, with most other models within ±2% of the range, such as Tree, LR and NN. RF sets a high misclassification frequency at 17.6%, but the version that appears to be the highest is the k-NN system with a 96.6% MSE.

From Fig. 5.2 and the parameters shown in Table 5.1, it was proposed that with a small number of k-NN neighbours, the identification rate is the highest relative to other models. This finding was regarded to be the most startling conclusion from the data collected in the earlier findings of the chapter in Fig. 4.8, where it was previously described that the SVM models achieve the highest precision relative to the other models under investigation. Miscellaneously, following the suggested approaches as discussed in this subchapter, the k-NN model shows substantially marginal differences in the classification rate compared to the alternate classifiers as investigated in Fig. 5.2. Table 5.2 provides a number of prototypes with a high accuracy rate in order to achieve a definitive evaluation.

Table 5.2 Test set error on selected models

		Test set							
		SVM		k-NN		RF		NN	
Error		17.32%		3.53%		7.66%		17.30%	
	True	H	S	H	S	H	S	H	S
Confusion matrix	H	5771	1777	6734	394	6518	867	2728	903
	S	1193	8403	375	9641	446	9313	432	3652

Table 5.3 MSE using the *k*-NN model on train set

Features		Train set					
		Two-dimension (PCA)		8 features (PCA varimax)		16 features	
Error		0.0% ±0.0%		0.0% ±0.0%		0.0% ±0.0%	
	True	H	S	H	S	H	S
Confusion matrix	H	21,040	0	21,106	0	21,106	0
	S	0	30,390	0	30,324	0	30,324

The table shows the proportions of MSE and the uncertainty matrixes of the main versions. The *k*-NN method categorizes the least error with an error of 3.53% preceded by an error of 7.66% for the RF system. The decision that can be made in relation to these two findings indicates that the level of identification for the hungry category is reasonably similar for 6734 and 6518 for the *k*-NN and RF, respectively. On the opposite, despite the equal misclassification error of 17.32 and 5.30% for SVM and NN, the incomparable frequency was diversely high as the SVM produced 5771 for the hungry class, while the NN generated 2728. Nevertheless, the phenomenon will be more noticeable as the MSE increases, and the frequency will be similar to that seen in the *k*-NN method (Table 5.3).

The test set classification of PCA shows a high variation in misclassification from 0.0 to 20.53% suggesting that this design is overrated where the test error margin is higher than the training error, while 8 features of PCA varimax and 16 features display a small error of 3.53% and 3.87%, respectively. From the data received, the parameters for all data sets are identical. The hyperparameter tuning analysis will now investigate the impact of the error rate on the validity array in general (Fig. 5.3).

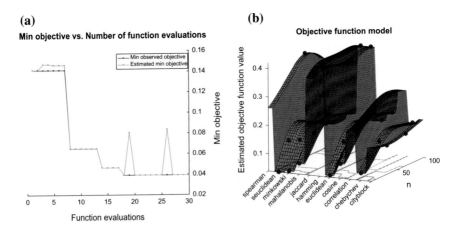

Fig. 5.3 Eight features of PCA varimax data set. **a** Function evaluation observed. **b** Objective functions evaluation observed

Table 5.4 MSE using the *k*-NN model on test set

Features		Test set					
		Two-dimension (PCA)		8 features (PCA varimax)		16 features	
Error		20.53%		3.53%		3.87%	
	True	H	S	H	S	H	S
Confusion matrix	H	2622	873	3330	154	3310	163
	S	887	4190	149	4939	169	4930

Large disparities were shown for the two-dimensional PCA derived classifier with a 14.82% error where the template was previously measured with a 0.0% error using the previous variable. Nonetheless, the change becomes noticeable when the failure falls from 14.59% relative to 20.53% as previously measured, as shown in Table 5.4. Likewise, performance fixes have been shown in every eight features of PCA varimax and 16 features of 2.94 and 2.97%. All data sets use the same parameters for a number of neighbours and standardize the Euclidean distance. Nonetheless, with respect to the minor gap variations in error rates, the latency of these two is relatively high, with 10 s dividing them between 15.59 s for eight features of the PCA varimax and 25.54 s for the 16 features This is ensuring that the eight features of the PCA varimax data sets offer the best solution for classifying hunger behaviour, with a validation range of 0.59% enhancement of the test set, as mentioned in Table 5.4 previously, resulting in a total accuracy of 97.06% (Tables 5.5 and 5.6).

Table 5.5 MSE using *k*-NN model on train set after hyperparameter tuning

Features		Train set (hyperparameter)					
		Two-dimension (PCA)		8 features (PCA varimax)		16 features	
Number of neighbours		100		1		1	
Distance metric		Euclidean		Standardize euclidean		Standardize euclidean	
Run-time (s)		11.97		15.59		25.54	
Error		14.82% ±0.001%		0.0% ±0.0%		0.0% ±0.0%	
	True	H	S	H	S	H	S
Confusion matrix	H	19,673	6253	21,106	0	21,106	0
	S	1367	24,137	0	30,324	0	30,324

Table 5.6 MSE using k-NN model on the validation set after hyperparameter tuning

Features		Validation set					
		Two-dimension (PCA)		8 features (PCA varimax)		16 features	
Error		14.59%		2.94%		2.97%	
	True	H	S	H	S	H	S
Confusion matrix	H	3288	1018	3352	119	3360	130
	S	233	4033	133	4968	125	4957

5.5 Summary

Through implementing hyperparameter tuning, this implies an improvement in the precision of classification on any data set. Classifiers evaluated different models to provide a better understanding of the correct template to be implemented. This is further explored by modifying model variables using hyperparameter optimization. The k-NN recorded the best reliability at 96.53% relative to the other models before the optimization approach was implemented. The findings shown after the introduction of the hyperparameter tuning of k-NN improved to 97.06%. In addition, the enactment of the model can be visible only with a single number of neighbours using a standardized Euclidean metric distance.

With these established methods and parameters, the processing performance should improve by increasing the storage consumption of the device. The client will have a clear insight into the programming of a new system that feeds fish on the basis of the parameters contained in this report. Nonetheless, further investigations are needed to reduce the computational period of learning of the model in order to provide a robust system for the further production of the request feeder.

References

1. Klein A, Falkner S, Bartels S, Hennig P, Hutter F (2016) Fast bayesian optimization of machine learning hyperparameters on large datasets. In: 20th international conference on artificial intelligence and statistics (AISTATS), JMLR, W&CP, vol 54. Fort Lauderdale, Florida, USA, 20–22 April 2017
2. Močkus J (1975) On bayesian methods for seeking the extremum. In: Proceedings of the IFIP technical conference. Springer-Verlag, Berlin, Heidelberg, pp 400–404
3. Shalev-Shwartz S, Ben-David S (2013) Understanding machine learning: from theory to algorithms
4. Robotham H, Castillo J, Bosch P, Perez-Kallens J (2011) A comparison of multi-class support vector machine and classification tree methods for hydroacoustic classification of fish-schools in Chile. Fish Res 111:170–176. https://doi.org/10.1016/J.FISHRES.2011.07.010
5. Pohar M, Blas M, Turk S (2004) Comparison of logistic regression and linear discriminant analysis: a simulation study. Metod Zv 1:143–161

6. Fransen BR, Duke SD, McWethy LG, Walter JK, Bilby RE (2006) A logistic regression model for predicting the upstream extent of fish occurrence based on geographical information systems data. North Am J Fish Manag 26:960–975. https://doi.org/10.1577/M04-187.1
7. Gurney K, York N (1997) An introduction to neural networks. Taylor & Francis, Inc., Bristol, PA, USA
8. Allken V, Handegard NO, Rosen S, Schreyeck T, Mahiout T, Malde K (2019) Fish species identification using a convolutional neural network trained on synthetic data. ICES J Mar Sci 76:342–349. https://doi.org/10.1093/icesjms/fsy147

Chapter 6
Concluding Remarks

Abstract The established automatic request feeder was able to understand the objective of this study through which to collect meaningful data without neglecting any vital information, including the feeding time, the feeder cause between the hungry and the happy, and the most relevant, extracting the notable features of the fish movement behaviours across the test. The contribution and future work shall be drawn up in conjunction with the goals reached in this report.

Keywords Classification · Fish hunger behaviour · Automated demand feeder · Contributions · Future work

6.1 Overview

The pre-processing stage allows the extraction features to be identified with the application of the DTW and decreases the sampling rate in order to minimize the computational complexity. The choice features put on this sample were deserved of the ML techniques, because they reflect the important characteristics that could enable a higher classification level. In contrast, by actively choosing the features depending on the varimax variable load rotation, the program will only need to pick eight features, as this can be seen by contrasting the PCA and 16 features. The chosen features of 0.5, 1.0, 1.5 and 2.0 min provide acceptable predictors for the classifier, particularly when using mean and variability for the features.

Through implementing hyperparameter tuning, this implies an improvement in the precision of classification on any dataset. Classifiers evaluated different models to provide a better understanding of the correct template to be implemented. This is further explored by modifying model variables using hyperparameter optimization. The k-NN recorded the best reliability at 96.53% relative to the other models before the optimization approach was implemented. The findings shown after the introduction of the hyperparameter tuning of k-NN improved to 97.06%. In fact, the application of the template can be visible only with a single number of neighbours using a fixed Euclidean metric length.

© The Author(s), under exclusive license to Springer Nature Singapore Pte Ltd. 2020   59
M. A. Mohd Razman et al., *Machine Learning in Aquaculture*,
SpringerBriefs in Applied Sciences and Technology,
https://doi.org/10.1007/978-981-15-2237-6_6

With these defined methods and parameters, the processing performance should improve by increasing the storage consumption of the device. The client will have a clear insight into the programming of a new system that feeds fish on the basis of the parameters contained in this report. Nonetheless, further studies are needed to reduce the computational period of learning of the model in order to provide a robust system for the further production of the request feeder.

6.2   Main Contributions

The main contribution of this research is the application of visualization methods and machine learning in the analysis of hunger behaviour. It concerns primarily on the convergence of marine biology and mechatronics technology with data analytics. To date, the other imaging studies recorded in the literature do not examine the circadian rhythm of the fish that is closely associated with the condition of hunger as in time series, in general through the use of an automatic feeder.

To date, the features derived as elaborated in the present study have yet to be examined, in general with regard to the identification of hunger states specifically in time-series classification. It has been shown in this experiment that the stated combination of function technology contributes to reasonably accurate identification of the starvation condition of *Lates calcarifer*. It is also worth noting that the different hyperparameters modified further increased the level of identification in the assessment of hunger behaviour.

6.3   Future Work

Under the pretext of enhancing the current research work, the following suggestions are proposed where, for example, the varying set-up of the camera, as well as the positioning of the sensor, could be examined using the suggested methodology. With the emergence of deep learning (DL), especially when it comes to image processing, DL can be used to obtain new features that are not possible by traditional ML techniques.

This research used a Bayesian optimization methodology to model hyperparameters. It is worth researching the utilization of other metaheuristic optimization algorithms, such as particle swarm optimization, genetic algorithm, to mention several in the determination of optimum hyperparameters. The future study should also look at the real-time application of the built ML template.

Printed in the United States
By Bookmasters

SpringerBriefs in Applied Sciences and Technology

Nanoscience and Nanotechnology

Series Editors

Hilmi Volkan Demir, Nanyang Technological University, Singapore, Singapore

Alexander O. Govorov, Clippinger Laboratories 251B, Department of Physics and Astronomy, Ohio University, Athens, OH, USA

Indexed by SCOPUS Nanoscience and nanotechnology offer means to assemble and study superstructures, composed of nanocomponents such as nanocrystals and biomolecules, exhibiting interesting unique properties. Also, nanoscience and nanotechnology enable ways to make and explore design-based artificial structures that do not exist in nature such as metamaterials and metasurfaces. Furthermore, nanoscience and nanotechnology allow us to make and understand tightly confined quasi-zero-dimensional to two-dimensional quantum structures such as nanopalettes and graphene with unique electronic structures. For example, today by using a biomolecular linker, one can assemble crystalline nanoparticles and nanowires into complex surfaces or composite structures with new electronic and optical properties. The unique properties of these superstructures result from the chemical composition and physical arrangement of such nanocomponents (e.g., semiconductor nanocrystals, metal nanoparticles, and biomolecules). Interactions between these elements (donor and acceptor) may further enhance such properties of the resulting hybrid superstructures. One of the important mechanisms is excitonics (enabled through energy transfer of exciton-exciton coupling) and another one is plasmonics (enabled by plasmon-exciton coupling). Also, in such nanoengineered structures, the light-material interactions at the nanoscale can be modified and enhanced, giving rise to nanophotonic effects.

These emerging topics of energy transfer, plasmonics, metastructuring and the like have now reached a level of wide-scale use and popularity that they are no longer the topics of a specialist, but now span the interests of all "end-users" of the new findings in these topics including those parties in biology, medicine, materials science and engineerings. Many technical books and reports have been published on individual topics in the specialized fields, and the existing literature have been typically written in a specialized manner for those in the field of interest (e.g., for only the physicists, only the chemists, etc.). However, currently there is no brief series available, which covers these topics in a way uniting all fields of interest including physics, chemistry, material science, biology, medicine, engineering, and the others.

The proposed new series in "Nanoscience and Nanotechnology" uniquely supports this cross-sectional platform spanning all of these fields. The proposed briefs series is intended to target a diverse readership and to serve as an important reference for both the specialized and general audience. This is not possible to achieve under the series of an engineering field (for example, electrical engineering) or under the series of a technical field (for example, physics and applied physics), which would have been very intimidating for biologists, medical doctors, materials scientists, etc.

The Briefs in NANOSCIENCE AND NANOTECHNOLOGY thus offers a great potential by itself, which will be interesting both for the specialists and the non-specialists.

More information about this subseries at http://www.springer.com/series/11713

T. Serkan Kasirga

Thermal Conductivity Measurements in Atomically Thin Materials and Devices

 Springer

T. Serkan Kasirga
UNAM-Institute of Materials
Science and Nanotechnology
Bilkent University
Ankara, Turkey

ISSN 2191-530X   ISSN 2191-5318   (electronic)
SpringerBriefs in Applied Sciences and Technology
ISSN 2196-1670   ISSN 2196-1689   (electronic)
Nanoscience and Nanotechnology
ISBN 978-981-15-5347-9   ISBN 978-981-15-5348-6   (eBook)
https://doi.org/10.1007/978-981-15-5348-6

This Springer imprint is published by the registered company Springer Nature Singapore Pte Ltd.
The registered company address is: 152 Beach Road, #21-01/04 Gateway East, Singapore 189721, Singapore

to my family

Preface

A thorough understanding of the thermal properties of materials may enable improved heat management in electronics and could pave the way for efficient energy scavenging from the waste heat in many processes. Conventional investigation of the thermal properties of materials has been focused on the engineering of grain boundaries in polycrystalline bulk samples to improve the thermoelectric materials figure of merit, zT, that is a measure of the efficiency of the thermoelectric conversion and it is a function of the Seebeck coefficient, thermal conductivity, electrical conductivity and temperature. The higher zT values mean higher efficiency in the thermoelectric conversion. An efficient thermoelectric material would ideally have high electrical conductivity and Seebeck coefficient and low thermal conductivity. However, typically these quantities are deeply related to each other. For instance, a material with high thermal conductivity according to the Wiedemann-Franz law has high electrical conductivity as well.

With the advent of the graphene in 2004, materials that are formed by van der Waals stacking of atomically thin two-dimensional (2D) layers have attracted a great deal of interest due to the potential applications in electronics and optoelectronics as various degrees of freedoms such as valley and spin can be tuned via gating and strain engineering. Moreover, via stacking of various layers of different materials, heterostructures with novel functionality can be created. This provides a whole new set of engineering materials in various fields of science and technology.

In this brief, first I will introduce the thermal properties of 2D materials. The focus, however, will be on the thermal conductivity measurement methods commonly used on 2D materials to extract the thermal properties. This brief is intended to be an overview of the available thermal conductivity measurement methods for the atomically thin materials. Due to its brief nature, many technical details will be referenced to the relevant literature. Thus, a certain level of understanding on the measurement methods is expected from the reader. Finally, a great part of the brief is allocated to the bolometric thermal conductivity measurement method introduced by our group recently. I discuss many points on the measurement technique in great detail in this brief that are not available in any other resources.

Ankara, Turkey   T. Serkan Kasırga
March, 2020

Acknowledgements

I gratefully acknowledge the support from Turkish Scientific and Technological Research Council (TÜBİTAK) under the grant no: 118F061. Also, I would like to thank Bilkent University National Nanotechnology Research Center, supported under the legislation 6550 for the infrastructure support that made all the research possible. Finally, I would like to thank my graduate students who were part of the above mentioned grant, especially Onur Çakıroğlu who conducted the experimental studies regarding the bolometric thermal conductivity measurement method.

Contents

About the Author

Dr. T. Serkan Kasırga received his Ph.D. from the University of Washington (Seattle, Washington) in 2013. Thereafter, he moved to Bilkent University—Institute of Materials Science and Nanotechnology as an assistant professor and he is the principal investigator of the Strongly Correlated Materials Research Laboratory. His research interests are focused on the solid-state phase transitions at various dimensions. He investigates the optical, mechanical, electrical and magnetic properties of correlated materials.

Acronyms

2D	Two dimensional
AFM	Atomic Force Microscopy
FEM	Finite Element Analysis
MBE	Molecular Beam Epitaxy
MOCVD	Metal-Organic Chemical Vapor Deposition
TCR	Temperature Coefficient of Resistance
TEM	Transmission Electron Microscopy
TMDC	Transition Metal Dichalcogenide

Chapter 1
Atomically Thin Materials

Abstract In this chapter, I will provide a brief overview of atomically thin materials that are formed by layers held together by van der Waals forces or weak covalent bonding. These materials provide a unique and cheap way of studying plethora of phenomena. Perhaps, the relative simplicity of the methods that are commonly used in the studies of two dimensional (2D) materials are one of the main reasons why they attracted attention at this level since the advent of graphene. After an introduction to the properties of 2D materials, I will talk about the methods to obtain 2D materials and conclude the chapter with the possibilities of heterostructures of 2D materials.

Keywords Atomically thin materials · Two dimensional materials · Graphene · Transition metal dichalcogenides

1.1 Introduction to Atomically Thin Materials

Materials that are formed by layers which are held together by van der Waals forces or weak covalent bonding can be mechanically or chemically exfoliated down to an in-plane covalently bonded single layer. The history of 2D materials dates back to 1960's. Graphene, one atom thick graphite layer, is among the first examples that has been isolated and studied in depth as a monolayer as early as 1980. However, the real debut of the 2D materials has been made by the seminal paper by Novoselov and Geim where they studied the graphene under electric and magnetic field [1]. The high quality of the crystals combined with the simplicity of obtaining them lead many researchers to ditch more complicated 2D electron gas materials for graphene. This is followed by demonstration of many effects in graphene. Further, other layered materials known to be mechanically exfoliable are joined the parade. As of 2020, more than a thousand materials have been identified as "easily" exfoliable and almost twice is identified as potentially exfoliable [2]. This overwhelming number of materials enables a plethora of novel physics as the number of layers often has a significant influence in the electrical, optical and thermal properties of the materials. As an instance, when the monolayer limit is approached, MoS_2 transitions from an indirect to a direct band gap with the change in the gap energy as well. Another

T. S. Kasirga, *Thermal Conductivity Measurements in Atomically
Thin Materials and Devices*, Nanoscience and Nanotechnology,
https://doi.org/10.1007/978-981-15-5348-6_1

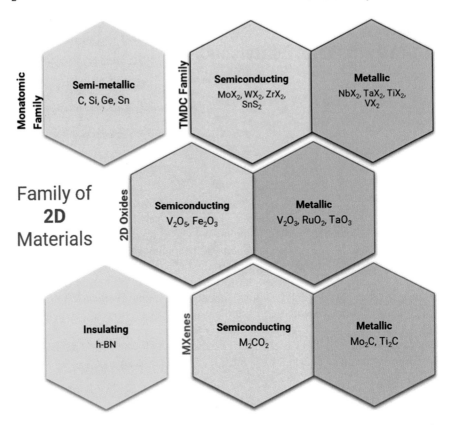

Fig. 1.1 A schematic showing the greater family of 2D materials. Although the great majority of the 2D materials that has been extensively studied so far belongs to the semiconductors of the TMDCs, there are still hundreds of 2D layers to be investigated

tuning parameter is the formation of the heterostructure of various materials. MoS_2 can form both in-plane and van der Waals heterostructures with another 2D material. As recently demonstrated, van der Waals heterostructure of bilayer of a material with a small twist shows distinctly different electrical properties due to formation of the Moiré potentials to modify the energy landscape of the charge carriers.

In terms of electrical conductivity there are insulating, semiconducting, semi-metallic and metallic 2D materials. Among the semiconductors there are famous examples such as MoS_2 or WS_2 (or their selenides) where they show direct band gap in the monolayer and indirect band gap from bilayer to bulk. The strong spin-orbit coupling with the broken inversion symmetry leads to optically accessible valley physics. Another commonly studied semiconducting 2D material has been the black phosphorous. It offers a thickness tunable direct bandgap with a high carrier mobility over $1000\,cm^2$/Verses. Despite being unstable in the ambient, the prospects it offers in applications have attracted a great deal of attention. Metallic 2D materials have started attracting interest as more and more studies showed high quality contact formation to the semiconducting 2D materials through the edge epitaxy. Moreover, many metallic 2D materials show intriguing properties. The exotic effects observed in the bulk such as superconductivity, charge density waves or Kondo physics typically gets modified as the material thickness approaches to the monolayer limit due to various reasons. As an instance $2H-TaS_2$ shows superconducting transition at an elevated temperature with the hints of 2D superconductivity in 6 layers [3]. It is also possible to obtain ferromagnetism in the 2D limit in materials such as CrI_3 and $Cr_2Ge_2Te_6$. This magnetism can be tuned with the number of layers as well as the external pressure on the sample.

The "periodic table" of 2D materials is yet to be completed and seems like it is going to offer a new "chemistry" that uses 2D layers rather than atoms (Fig. 1.1). There are many good literature reviews on the topic and as I would like to focus on the thermal properties of 2D materials on this brief, I will stop the introduction to the 2D materials here and refer the reader to these excellent reviews for further information.

1.2 Obtaining 2D Materials

2D materials can be obtained via various methods. The most commonly employed method is the mechanical exfoliation of atomically thin crystals from the bulk using a sticky tape [1]. This produces very high-quality crystals; however, the yield of the monolayer crystals is sufficient only for proof-of-concept demonstrations. Another commonly used method is the chemical vapor deposition (CVD) of the 2D flakes. This method typically yields lower quality samples as compared to the mechanical exfoliation yet results in large crystals in multiples of numbers. By controlling the growth parameters, it is also possible to obtain wafer large polycrystalline mostly monolayer films. Further improvement in uniformity of the monolayers in CVD can be achieved in metal-organic deposition of the monolayers on substrates like sapphire. However, as demonstrated in many studies dislocations and defects in the grain boundaries in polycrystalline films are prone to oxidation and vastly alters the overall sample properties. Finally, liquid exfoliation is another common method to obtain the monolayers in liquid via intercalating ions in between the layers. The method produces sub-micron sized crystals and not suitable for many applications and proof-of-concept demonstrations.

1.2.1 Mechanical Exfoliation of 2D Materials

Mechanical exfoliation of the 2D materials has been a revolutionary method to investigate atomically thin layers. First of all, it only requires the bulk crystal and a sticky tape. As the intralayer forces are very week, mechanical tape can easily separate the layers from each other. These layers can be deposited on smooth surfaces with

Fig. 1.2 **a** Tape with the 2D layers exfoliated from the bulk and the oxidized silicon chip is shown. **b** and **c** shows the oxygen plasma treatment of the substrate to increase the adhesion of the monolayers and heated the substrate is heated to remove any water residues. **d** Adhesion of the tape on the substrate to deposit the monolayers is photographed and **e** the substrate after mechanical exfoliation. **f** Optical microscope micrograph of the substrate shows large layered material. Reprinted with the permission from Huang et al. [4] Copyright (2015) American Chemical Society

enough attraction by sticking the tape. This produces crystals of varying thicknesses. The yield heavily depends on the substrate, stickiness of the tape and the type of the crystal to be deposited. Moreover, oxidation of the surface further effects the yield of the atomically thin layers to be deposited (Fig. 1.2). There are recent efforts that results in reproducible exfoliation [4] of the monolayers up to in centimeter size [5]. As an example gold evaporated stacks of polyvinylpyrrolidone (PVP) with thermal release tape can pick a single monolayer from the bulk crystal and can be deposited on a surface with a subsequent etching of the gold on the top. Type of the tape used is also important in improving the yield. Typically, Nitto brand tape with very low adhesion is used for exfoliation of many materials yet some others require a stickier tape for a higher yield.

Currently, most of the leading laboratories on 2D materials around the world are now performing exfoliation within an inert chamber as even the most oxidation durable materials are adversely affected by the oxygen and the moisture in the ambient. Recent developments in the oxygen and moisture free chambers will most likely to make the process more efficient and cheaper [6]. In particular, the booming interest on the physics of the twisted multi-layer structures namely the Moire superlattices increased the interest in ultra-clean interfaces between the layers [7, 8]. Interlayer contamination leads to significant reduction in the electronics of the superlattice.

The sticky tape exfoliation of the 2D layered materials begin with the bulk crystal either mined (typically for graphite or MoS_2) or synthesized via chemical vapor transport (CVT) synthesis method. Most of the CVT based synthesis of the bulk material follows a similar recipe. Precursor materials such as the metal and the chalcogens mixed with a transport agent such as iodine are placed at the opposing sides of a vacuum sealed quartz ampule. Then, a long growth procedure starts with well defined temperature steps. A few millimeter large crystal may take up to a month to form at elevated temperatures. There are several companies that provide the bulk materials such as 2d Semiconductors Inc. (USA) and HQ Graphene (the Netherlands). However, most bulk crystals of 2D materials are notoriously difficult to obtain in the monolayer.

1.2.2 Chemical Vapor Deposition of Monolayers

Chemical vapor deposition of monolayer materials begs for an entire book, perhaps several volumes of books. Chemical vapor deposition can be performed in several ways. First of all, it is possible to perform the growth using a vapor-vapor-solid mechanism starting with solid precursors and forming gasses. For instance, MoS_2 monolayers can be synthesized using MoO_3 and Se precursors with H_2 gas on a substrate. It has been shown that if a salt such as NaCl or KI is added to the metal oxide precursor, due to formation of more volatile gaseous and liquid precursors, the nucleation and the growth of the crystals become more controllable. This also enables the synthesis of layered metalchalcogenides from metal precursors with little volatility. It is also possible to use the metalorganic chemical vapor deposition method for wafer-scale synthesis of monolayers. However, the quality of the film is still not at the desired level due to the small grain sizes. Also, the cost of the synthesis is still not justifiable for many applications for the quality of the crystals obtained (Fig. 1.3).

Finally, I would like to highlight one of the methods we introduced [9]. Using a custom-made chemical vapor deposition chamber with an optical observation window, we enabled the real-time control of the monolayer synthesis. This enables the fabrication of unique heterostructures of 2D materials as well as establishing the required parameters for the crystal growth. The chamber is a cold-walled chamber that is composed of discrete heaters that are designated for heating different precur-

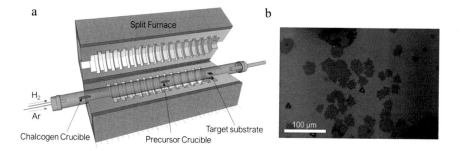

Fig. 1.3 **a** Schematic of a typical CVD split-tube furnace for 2D materials growth is shown. **b** A representative image of WSe₂ monolayers shown

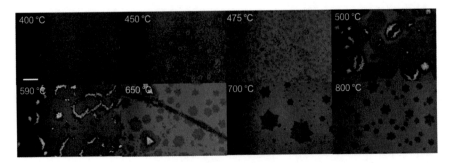

Fig. 1.4 Figure showing the results of the substrate temperature on the nucleation of MoSe₂ crystals on oxidized silicon substrates. This is different than what other studies reported as in conventional CVD chamber it is not possible to change the substrate temperature without creating thermal gradients over the sample or changing the precursor temperatures. Scale bar is 20 μm

sors and the substrate that is directly located under an optical window. With an high magnification microscope equipped with 40× objective (ultra-long working distance with cover slip compensation up to 2 mm), high resolution tracking of the growth in-real time is possible. Moreover all the gas flow parameters are controlled via software that enables accurate timing of the reaction gasses. In our paper, we demonstrated the relevant mechanisms for the salt assisted synthesis of MoSe₂, WSe₂ monolayers and their lateral and vertical heterostructures in a single-step. We also characterized the growth rate via vapor-solid-solid and vapor-liquid-solid mechanisms. Moreover, with the chamber we studied the real effect of the substrate temperature on the crystal formation. Unpublished results are shown in Fig. 1.4 that shows the sole effect of the substrate temperature on the nucleation of the MoS₂ crystals with all the other parameters are kept constant.

1.2.3 Heterostructures of 2D Materials

One of the marvels of the modern technology is the layer-by-layer deposition of semiconducting layers with varying band gaps. The deposition of the layers mostly uses molecular beam epitaxy, metalorganic chemical vapor deposition of pulsed laser deposition. This is a research field on its own and it has been a technological break-through. However, deposition of these semiconducting layers is extremely costly due to high purity and ultra-high vacuum conditions required for the deposition of the layers. Similar heterostructures can be fabricated by using 2D materials as well. It is possible to transfer both mechanically exfoliated and CVD grown crystals on various substrates and various other 2D crystals. New functionality can be gained through fabrication of such hetero-structures due to the electronic and phononic intralayer coupling (Fig. 1.5). Unlike the synthesized heterostructures, there is also so-called Moiré degree of freedom that results in novel electronic band formation [7]. The fabrication procedure of the heterostructures are rather straightforward however requires an ultra-clean environment for the best results. Gant et al. [6] recently published a very comprehensive paper on high quality transfer of 2D materials. I would like to refer the reader to their paper and the reference therein for a more detailed information.

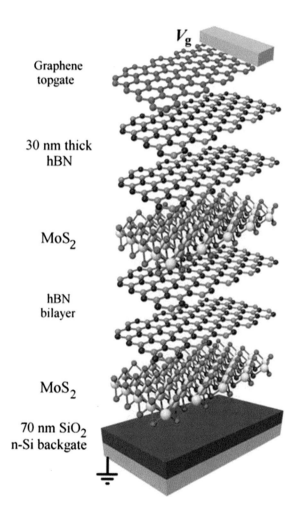

Fig. 1.5 A schematic showing the stacking of the van der Waals heterostructure of 2D layers. Reprinted with the permission from Calman et al. [10]. Copyright (2016) AIP Publishing

V_g

Graphene topgate

30 nm thick hBN

MoS$_2$

hBN bilayer

MoS$_2$

70 nm SiO$_2$ n-Si backgate

1.3   A Suitable Platform to Study Fundamental Science and Explore Applications

In the previous sections I tried to give a brief overview of a wast field. I would like to conclude this chapter by discussing why 2D materials has been so popular in the last decade and a half. Creating non-existing features in solid-state systems require fabrication of carefully designed structures, called superlattices (SL), to be formed in very high quality layers. Most commonly employed method is layer by layer formation of these layers using molecular beam epitaxy (MBE). Operation of the MBE is extremely costly and the design of the structures require many iterations. The main reason is the lattice mismatch induced stress on the alternating layers. Thus buffer layers have to be investigated carefully to minimize such stresses to reduce the defect formation for high performance applications. Metal-organic chemical vapor deposition (MOCVD) is another commonly utilized method that is relatively cheaper to operate and can be scaled to multiple wafer deposition for industrial processes. However, the defects and impurities in MOCVD grown SLs are much more significant compared to MBE grown SLs. Pulsed laser deposition (PLD) is another method that can be used for evaporation of materials in a controlled way on the target substrate. However, PLD method is somewhat limited in the industrial scale applications and creation of the SLs. Moreover, just like MBE, MOCVD and PLD are very expensive equipment for a research group to have and operate. These reasons form the basis of the boom behind the 2D materials research.

As mentioned in the introduction, 2D materials form a huge family. More than hundreds of 2D materials have been isolated so far. It is possible to form the heterostructures of these 2D materials to create novel SLs. The major advantage over other SL forming methods is that the 2D crystals can be mechanically exfoliated and at the proof of concept level it is possible to study the basic physics and applications of the SLs. Furthermore, there are new degrees of freedoms that can be employed to gain novel functionality in the SL structure. For instance formation of Moire patterns due to slight rotational angles among the layers results in novel properties in the 2D SL structures. The entry barrier to the field is very low: crystals can be exfoliated from the bulk material using a sticky tape and can be deposited on any suitable substrate. Even the formation of the heterostructures can be done with very high precision at a very low initial and operational cost. This allows many researchers around the world to easily adopt the required methods to study the 2D materials. This is a positive feedback loop. As more researchers enter the field, it becomes more visible and as it becomes more visible more researchers keep entering the field. Still, there are many materials waiting to be discovered with many phenomena waiting to be explored.

1.3.1 Applications of 2D Materials

Being a relatively young research field, it is not realistic to expect applications that can compete with concurrent technologies. However, the prospects of applications are quite tantalizing and there are many research groups working on finding ways to make 2D materials commercially viable. This requires high quality materials to be synthesized in wafer scale and possibly the removal of the 2D layer from the growth substrate to form flexible optoelectronics and heterostructures. The major advantage of the 2D materials will become more apparent in the long run with the use of novel degrees of freedom available in these materials. As an instance, valley polarization that exists in inversion symmetry broken semiconducting monolayer provides a way to selectively couple circularly polarized light to the electronic states. Further examples can be given in the field of 2D magnets. These 2D magnets can be controlled depending on the number of layers.

As the focus of this brief is on the thermal conductivity measurements of 2D materials, I would like to conclude this chapter by providing an introduction on the thermal properties of 2D materials. Single crystal graphene flakes are shown to demonstrate very high thermal conductivity in the range of 1450–5300 W/m K [11–13]. Possibility of having such high thermal conductivity values is very promising for heat dissipation applications. Another exciting 2D material is the h-BN. Despite being an insulator, it exhibits considerably high thermal conductivity in the vicinity of 300 W/m K [14–16]. To put in a context, thermal conductivity of Si at room temperature is around 130 W/m K and even for copper and silver it is around 400 W/m K. These high thermal conductivity finds its particular use in heat dissipation in electronics and cooling applications. Another particularly important side is the ultra-low thermal conductivity measurements for heat shielding and high ZT materials for thermoelectric harvesting. However, as mentioned in the first sections, 2D materials family is a large one and there are many different types of 2D materials that are not measured to date.

References

1. Novoselov KS (2004) Science 306(5696):666. https://doi.org/10.1126/science.1102896. https://www.sciencemag.org/lookup/doi/10.1126/science.1102896
2. Mounet N, Gibertini M, Schwaller P, Campi D, Merkys A, Marrazzo A, Sohier T, Castelli IE, Cepellotti A, Pizzi G, Marzari N (2018) Nat Nanotechnol 13(3):246. https://doi.org/10.1038/s41565-017-0035-5.
3. Navarro-Moratalla E, Island JO, Manãs-Valero S, Pinilla-Cienfuegos E, Castellanos-Gomez A, Quereda J, Rubio-Bollinger G, Chirolli L, Silva-Guillén JA, Agraït N, Steele GA, Guinea F, Van Der Zant HS, Coronado E (2016) Nat Commun 7:1. https://doi.org/10.1038/ncomms11043
4. Huang Y, Sutter E, Shi NN, Zheng J, Yang T, Englund D, Gao HJ, Sutter P (2015) ACS Nano 9(11):10612. https://doi.org/10.1021/acsnano.5b04258
5. Liu F, Wu W, Bai Y, Chae SH, Li Q, Wang J, Hone J, Zhu XY (2020) To be published 906(February):1

6. Gant P, Carrascoso F, Zhao Q, Ryu YK, Seitz M, Prins F, Frisenda R, Castellanos-Gomez A (2020) 2D Materials (2020). https://doi.org/10.1088/2053-1583/ab72d6
7. Cao Y, Fatemi V, Fang S, Watanabe K, Taniguchi T, Kaxiras E, Jarillo-Herrero P (2018) Nature 556(7699):43. https://doi.org/10.1038/nature26160
8. Cao Y, Fatemi V, Demir A, Fang S, Tomarken SL, Luo JY, Sanchez-Yamagishi JD, Watanabe K, Taniguchi T, Kaxiras E, Ashoori RC, Jarillo-Herrero P (2018) Nature 556(7699):80. https://doi.org/10.1038/nature26154
9. Rasouli HR, Mehmood N, Çakıroğlu O, Kasırga TS (2019) Nanoscale 11(15):7317. https://doi.org/10.1039/C9NR00614A. http://pubs.rsc.org/en/Content/ArticleLanding/2019/NR/C9NR00614A, http://xlink.rsc.org/?DOI=C9NR00614A
10. Calman EV, Dorow CJ, Fogler MM, Butov LV, Hu S, Mishchenko A, Geim AK (2016) Appl Phys Lett 108(10):101901. https://doi.org/10.1063/1.4943204
11. Balandin AA, Ghosh S, Bao W, Calizo I, Teweldebrhan D, Miao F, Lau CN (2008) Nano Lett 8(3):902. https://doi.org/10.1021/nl0731872
12. Chen S, Moore AL, Cai W, Suk JW, An J, Mishra C, Amos C, Magnuson CW, Kang J, Shi L, Ruoff RS (2011) ACS Nano 5(1):321. https://doi.org/10.1021/nn102915x
13. Cai W, Moore AL, Zhu Y, Li X, Chen S, Shi L, Ruoff RS (2010) Nano Lett 10(5):1645. https://doi.org/10.1021/nl9041966
14. Jo I, Pettes MT, Kim J, Watanabe K, Taniguchi T, Yao Z, Shi L (2013) Nano Lett 13(2):550. https://doi.org/10.1021/nl304060g
15. Zhou H, Zhu J, Liu Z, Yan Z, Fan X, Lin J, Wang G, Yan Q, Yu T, Ajayan PM, Tour JM (2014) Nano Res 7(8):1232. https://doi.org/10.1007/s12274-014-0486-z
16. Cai Q, Scullion D, Gan W, Falin A, Zhang S, Watanabe K, Taniguchi T, Chen Y, Santos EJ, Li LH (2019) Sci Adv 5(6):1. https://doi.org/10.1126/sciadv.aav0129

Chapter 2
Thermal Conductivity Measurements in 2D Materials

Abstract Measuring the thermal conductivity of materials is a very important field as the continuation of the improvement in modern electronics and optoelectronics heavily depends on the thermal management. Both high thermal conductivity and low thermal conductivity materials are required in the device design. Besides the fields mentioned above, excess heat scavenging via thermoelectric devices is an ever-growing field. Thermoelectric devices performance is determined by the figure of merit $Z = S^2 \sigma \kappa$ where S is the Seebeck coefficient, σ is the electrical conductivity and κ is the thermal conductivity. Materials that are good electrical conductors and thermal insulators are needed for efficient thermoelectric devices.

Keywords Thermal conductivity theory · Thermal conductivity measurement methods · Raman thermometry · Micro-bridge thermometry · Time-domain thermoreflectance · Bolometric thermal conductivity measurements

2.1 Introduction to Thermal Conductivity and Thermal Conductivity Measurement Techniques

After the advent of the graphene in 2004, the study of the solid-state phenomena is revolutionized as the atomically thin materials become widely available. Mechanically exfoliated monolayers of many two-dimensional materials revealed plethora of phenomena and enabled the possibility of novel electronics and optoelectronics applications due to the new electronic degrees of freedom that can be relatively easily studied. A further dimension in the study of the atomically thin materials has been introduced via stacking of various monolayers to achieve new functionality. All these developments in the field have led to the proof-of-concept demonstrations at the prototype level. Although, there is not enough incentive for the industry to adopt these newly discovered materials yet, with the recent developments in the large area high quality synthesis of the 2D materials within a decade or so they most likely compliment the engineering materials. An integral part of this process is the measurement of the thermal conductivity values in 2D materials starting from a few monolayers down to the single layer limit as well as their heterostructures to aid the engineering of the high-performance devices.

T. S. Kasirga, *Thermal Conductivity Measurements in Atomically Thin Materials and Devices*, Nanoscience and Nanotechnology, https://doi.org/10.1007/978-981-15-5348-6_2

The challenge in measuring the thermal conductivity in atomically thin materials is self-evident. As we will discuss in the following parts, thermal conductivity measurements in bulk materials and even in thin films is relatively easier due to the large lateral sizes. At its current stage, most 2D materials can only be obtained as crystals of a few tens of micrometers in-plane direction. This challenge in handling of the samples complicated the measurement methods and decreases the accuracy of the temperature profiling of the samples as well as the heat measurements. Thus, many measurement techniques fail to measure the thermal conductivity precisely. In this chapter, I will introduce thermal conductivity in solid media and common methods used to measure thermal conductivity. Before proceeding with the methods, I would like to provide a brief introduction to thermal conduction in solid media.

2.2   Thermal Conductivity in Solid Media

Thermal conductivity is a measure of the ability to transfer heat spatially in a material. The heat flux from the hotter side of the material to its colder side is proportional to the temperature gradient over a material through the negative of the thermal conductivity by the Fourier's law. A general formulation of the Fourier's law can be written as:

$$\mathbf{Q}(\mathbf{r}, t) = -\kappa \nabla T(\mathbf{r}, t) \tag{2.1}$$

where $\mathbf{Q}(\mathbf{r}, t)$ is the heat flux density vector along the different axes of the material, κ is the second-rank thermal conductivity tensor, $T(\mathbf{r}, t)$ is temperature gradient as a function of position, $\mathbf{r}$, and time, t. Heat in solids can be transferred via electrons and lattice vibrations (phonons). For the sake of theoretical simplicity electronic heat conduction can be studies as the heat conduction via electrons and holes, spin waves and other excitations. Plasmonic oscillations can also be an important heat transport mechanism in solids. Total thermal conductivity of the material is the algebraic sum of all the thermal conductivity contributions from various mechanisms . The electronic thermal conductivity can be studied using nearly free electron in a periodic potential. This treatment leads to the proof of the Wiedemann-Franz law:

$$L_0 = \frac{\kappa_e}{\sigma T} = 2.4452 \times 10^{-8} \ \text{W}\Omega/\text{K}^2 \tag{2.2}$$

Here, σ is the electrical conductivity of the material and the numerical value of the Lorentz number L_0 can be calculated in terms of the Boltzmann constant and the electronic charge. The Wiedemann-Franz law typically holds at high temperatures where the inelastic scattering of the charge carriers is relatively insignificant. Pure solid metals such as gold and silver exhibit ignoreable phonon thermal conductivity as the free electrons dominate the thermal conductivity. As we will discuss in the following sections, this is often not the case for the 2D metallic compounds as well as some transition metals and alloys. For the case of semiconducting materials, a similar expression can be derived based on equilibrium distribution of the electrons and holes. Theoretical treatment of the lattice contribution to the thermal conductivity requires addition of the effects of impurity scattering, isotope effects, etc. However, typically for materials with moderate impurity concentrations the dominant phonon assumption explains the phonon scattering contribution to the thermal conductivity adequately. Of course, discussions for the electronic and phononic thermal conductivity assumes that there are no correlations between the electrons and phonons. This is not entirely true for most of the transition-metal dichalcogenides that we will study in the following sections where electron-phonon correlations lead to intriguing phenomena such as charge density wave transitions. Phonons in a lattice can scatter due to several mechanisms such as phonon-phonon scattering, phonon impurity scattering, phonon-electron scattering and phonon-boundary scattering.

2.2.1 Phonon-Phonon Scattering

Phonon-phonon scattering mechanism can be classified under two categories: the normal process (N-process) that conserves the momentum after scattering and the umklapp (turn-over) process (U-process) that does not conserve the momentum. The U-process is a result of scattering that produces a wave-vector that is larger than the Brillouin zone of the crystal, thus falling into the opposite corner of the Brillouin zone. This literally flips over the momentum of the phonon after the scattering event and as it has ω^2 frequency dependence, it is much more dominant than the N-process with ω dependence at high temperatures for low-defect crystals. The relaxation time, τ_U, for the U-process can be written as:

$$\frac{1}{\tau_U} = 2\gamma^2 \frac{k_B T}{\mu V_0} \frac{\omega^2}{\omega_D} \tag{2.3}$$

where, ω_D is the Debye frequency, γ is the Gruneisen parameter, μ is the shear modulus, V_0 is the volume per atom and ω is the phonon frequency.

Another important thermal relaxation mechanism for low dimensional materials is the boundary scattering. Scattering of phonons from the boundaries of the material leads to thermal resistance when the boundaries are rough. The thermal mean free path, ℓ_B can be written as

$$\ell_B = D \frac{1+p}{1-p} \tag{2.4}$$

where D is the diameter of the sample and p is the specularity parameter. When $p = 1$, the scattering is perfectly specular while $p = 0$ it is perfectly diffusive.

As most of the metallic 2D materials show strong electron-phonon coupling, electron-phonon scattering should also be considered for another thermal resistance mechanism.

$$\frac{1}{\tau_{ph-e}} = \frac{n_e \epsilon^2 \omega}{\rho V^2 k_B T} \sqrt{\frac{\pi m^* V^2}{2 k_B T}} \exp\left(-\frac{m^* V^2}{2 k_B T}\right) \tag{2.5}$$

where n_e is the density of conduction electrons, ϵ is deformation potential, ρ is mass density and $m*$ is effective electron mass.

All the contributions to the thermal relaxation can be added with the Mattheissen's rule as the following:

$$\frac{1}{\tau_{Tot}} = \frac{1}{\tau_U} + \frac{1}{\tau_{ph-e}} + \frac{1}{\tau_B} \tag{2.6}$$

and using the kinetic theory of gasses, it is possible to write a relation between the thermal conductivity and the specific heat, C, phonon group velocity, v, and the thermal relaxation time, τ_{Tot}:

$$\kappa = \frac{1}{3}Cv^2\tau_{Tot} \tag{2.7}$$

Depending on the major heat carriers in the material, the specific heat can be electronic or the lattice.

There are many factors that influence the thermal conduction in materials. Temperature, crystallinity, impurities, defects, carrier density, electron-electron and electron-phonon correlations are a few of these factors. We will discuss the effect of each factor in detail as we discuss specific material families in the following sections. To conclude, I think it is necessary to add that there are various theoretical methods that can be used to obtain the thermal conductivity values such as molecular dynamics simulations besides the analytical approaches we introduced.

2.3 Thermal Conductivity of 2D Materials

2D layered materials exhibit a unique crystal structure that leads to high thermal anisotropy ranging from 50 to 300 slower heat conduction in cross-plane than in-plane [1–3]. The atoms in each layer are covalently bonded and the layers are stacked via van der Waals interactions. This results in slower heat transport across the layers as compared to the in-plane. Although most of the studies on thermal conductivity of 2D materials are theoretical, there is a good understanding of how heat is carried in these systems both in the monolayer limit as well as in the bulk. In the following subsections I summarize the recent understanding of thermal conduction in 2D materials. I constrained myself to single crystal, single domain properties of each 2D materials family as the composites made up of 2D materials as well as the polycrystalline samples opens up a whole new world for the thermal conductivities. Moreover, conventional methods for such composite and bulk film/crystalline materials are applicable. I limited my attention in this brief to more academic exploration of the 2D materials' thermal properties.

2.3.1 Thermal Conductivity of Graphene

Theory of the thermal conductivity of graphene has been extensively studied in the literature. Thermal conductivity of graphene is mostly due to the phonons as the electronic contribution at room temperature is about 10% of graphene when impurities are excluded. When impurities are included the first-principles calculations, the electronic contribution decreases to 0.5–8 % of the monolayer [4]. Phonons in the graphene can be classified as in-plane and out-of-plane (flexural) modes. Linear transverse acoustic (LA) and longitudinal acoustic (LA) modes are in-plane modes. Flexural modes are acoustic (ZA) and optical (ZO). Seol et al. [5] argued that ZA phonon modes carry most of the heat in graphene due to the large density of ZA flexural phonons resulting from their quadratic dispersion and to a selection rule for three-phonon scattering that arises from the reflection symmetry perpendicular to the graphene plane based on their experimental data. In a follow-up study, using the linearized Boltzmann transport equation Lindsay et al. [6] showed that despite the commonly accepted view, flexural acoustic phonons account for more that 70% of the total thermal conductivity [7]. Thus the choice of substrate is very important (see Fig. 2.1). This is also similar in h-BN [7]. For the cross-plane thermal transport a key quantity is the phonon mean free path. Molecular dynamics simulations and experimental measurements of the cross-plane thermal transport measurements showed that the phonon mean free path can be on the order of hundreds of nanometers [8–10]. Another important factor in the thermal conductivity of graphene is the isotope effect [11].

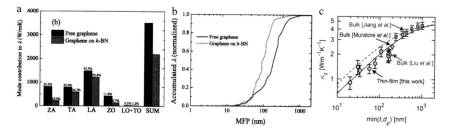

Fig. 2.1 a. Mode thermal conductivity of free and supported graphene calculated using molecular dynamics simulations showing the effect of substrate on the thermal properties. **b.** Figure shows the accumulative thermal conductivity as a function of the phonon mean free path in graphene and graphene on h-BN from molecular dynamics simulations. The figure is reprinted with the permission from Zou et al. [12]. Copyright (2017) AIP Publishing. **c.** Cross-plane thermal conductivity of MoS$_2$ as a function of thickness. The figure is reprinted with the permission from Sood et al. [13]. Copyright (2019) American Chemical Society

2.3.2 Thermal Conductivity of TMDCs

MoS_2 has been a prototypical material for TMDC studies due to ease of synthesis and high quality mined crystals. Thus, most thermal conductivity studies of TMDCs are focused on MoS_2. However, both experimentally and computationally there are large discrepancies in the thermal conductivity values reported on MoS_2. In-plane thermal conductivity values are in the range of 44–52 $Wm^{-1}K^{-1}$ [14, 15]. Of course the discrepancy might arise from the differences in the crystal thicknesses. Cross-plane thermal conductivity is also strongly effected by the thickness of the crystals due to very long phonon mean free paths along the z-axis [13]. From 20 to 1000 nm cross-plane thermal conductivity increases from 1 to 4 $Wm^{-1}K^{-1}$ as shown in Fig. 2.1 [3, 13, 16, 17].

Thermal conductivity can be greatly modified by the isotope effect similar to graphene. As an instance, it has been demonstrated that the thermal conductivity of MoS_2 prepared with ^{100}Mo isotope shows 50% enhanced thermal conductivity compared with the MoS_2 prepared with naturally occurring Mo isotopes [18]. This enhancement is attributed to the combined effects of reduced isotopic disorder and a reduction in defect-related scattering mechanisms.

2.3.3 Thermal Conductivity of Heterostructures

The heterostructures of 2D materials are very promising for enabling new applications using features that are not existent in the pristine materials [19]. There has been limited interest in the thermal conductivity of the heterostructures. Gao et al. theoretically studied the graphene-MoS_2 heterostructure using nonequilibrium molecular dynamics simulations and showed that the heterostructure has a lower thermal conductivity that decreases further with strain than graphene but higher than MoS_2 monolayer [20]. A similar feature is also demonstrated in perfect $MoS_2/MoSe_2$ heterostructures [21]. Beyond these studies, as discussed before, graphene on h-BN has been studied to a certain extend. There is still a lot to be done on the heterostructures of 2D materials.

2.4 Thermal Conductivity Measurement Methods in 2D Materials

Thermal conductivity measurements requires the measurement of heat transferred over the sample and measuring temperature is the most straightforward way to extract this information. However, measuring the temperature is not always straightforward. The problem is two-fold. First, if the sample size is very small, measuring the temperature requires a sensitive senor to probe the temperature, which complicates the device fabrication significantly. Second, as it is nearly impossible to fully isolate a material from its surrounding thermally, the heat transferred through other paths also has to be taken into the account. As the focus in this brief is on 2D materials, I would like to discuss the most commonly used methods on 2D materials.

Thermal conductivity measurement methods can be classified under two sections depending on the thermal equilibrium of the sample under measurement: transient methods and steady-state methods. In transient methods, thermal conductivity is measured in time (or frequency) domain when the thermal equilibrium has not reached with the surroundings. Most commonly used transient methods are the time-domain thermoreflectance spectroscopy and the 3ω method. In the steady-state methods, thermal equilibrium is established for the measurements. Most commonly used steady-state methods are the Raman thermometry based thermal conductivity measurement method, micro-bridge method. The method we introduced recently based on the bolometric effect is also a steady-state method. We will discuss all these methods in detail with their applicability in 2D materials.

2.4.1 Raman Thermometry

Raman spectroscopy is a versatile tool for materials characterization as the specific vibrational Raman active modes provide valuable information about the material. Some of the Raman active modes are sensitive to external influences such as strain, temperature and doping of the material. These external influences modify the stiffness among the constituent elements forming the material, thus leading to an observable shifts in the Raman spectrum. Raman thermometry is based on the temperature dependent shifts of the Raman active shifts. Thus, under sufficiently low laser illumination, it is possible to use the Raman shift as a thermometer [22, 23]. Although the method has been proposed to be used on silicon, it became very popular after being used to measure the thermal conductivity of graphene [24].

To be used in 2D materials, substrate preparation is mandatory. A substrate with circular holes are etched on the substrate and the nanosheet is transferred over the holes to suspend a certain section of the crystal. Then, a diffraction-limited laser spot is positioned over the center of the hole to measure the Raman spectrum at the lowest possible laser power at different substrate temperatures, T_{sub}. This provides the first-order temperature coefficient of Raman peak shift, $\chi_T = \Delta\omega/(T_2 - T_1)$. Typically, this parameter has to be measured over a very wide temperature range to get an accurate slope for χ_T. For instance, the plot shows a typical data set for MoS_2 adapted from Sahoo et al. [14], Yan et al. [25] and Zhang et al. [26] shows that a reliable slope can be extracted over a temperature range of 200 K. When the slope is measured in a narrower temperature range, the uncertainty in the slope becomes so large that the measured thermal conductivity value ranges significantly. Once χ_T is measured, the next step is the measurement of χ_P that is the Raman peak shift with the absorbed laser power. First of all, absorbance for the material under investigation has to be carefully measured. It should be also noted that the absorption can vary with

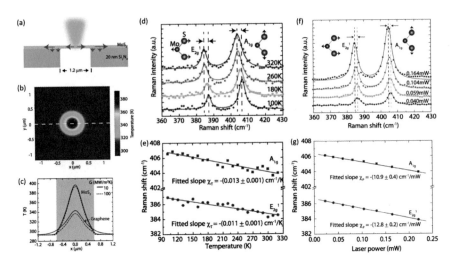

Fig. 2.2 a. Schematic of the side view of an MoS$_2$ crystal suspended over a hole with laser illumination from the top. Thermal simulation showing the heat distribution over the suspended and the supported parts of the crystal. **c.** Comparison of thermal distribution of graphene and MoS$_2$ monolayer under the same laser illumination at different thermal boundary conductance values. **d.** Raman peak shifts at various temperatures shown as the full spectrum and e. peak position shifts at various temperatures as a plot. **e.** Raman peak shifts as a function of the laser power shown as the full spectrum and **f.** peak shits at various powers. All the figures are reprinted with the permission from Yan et al. [25]. Copyright (2014) American Chemical Society

the temperature. Another consideration is the fact that the laser spot has a Gaussian power distribution, thus the local temperature distribution has to be weighted by the Gaussian profile of the laser spot. Thus, with the known mean temperature under the laser spot, the heat equation for the suspended and the supported part can be solved to give the thermal conductivity. The method is employed on various thin layers such as MoSe$_2$ [26, 27], WS$_2$ [28] (Fig. 2.2)

The method stands out as an exceptionally easy to implement method among the other thermal conductivity measurement methods that are applicable to 2D materials. This is mostly due to the simplicity of device fabrication and the experimental setup. Also, the analysis of the obtained data is much more simpler compared to the TDTR or 3ω methods. However, there are serious limitations of the method as well.

2.4.1.1   Limitations of the Raman Thermometry

Based on the first-principles calculations Vallabhaneni et al. [29] showed that different phonon polarizations are not in thermal equilibrium in graphene. This is the the well-accepted assumption of the Raman method. As a result, the method underestimates the thermal conductivity by a factor of 1.35–2.6. Further, they claim that such an underestimation is expected for other 2D materials when the optical-acoustic phonon coupling is weak.

Another limitation of the method is the requirement of having a Raman peak that is insensitive to the pre-tension due to suspension of the crystals over the hole. As it is very well known from the indentation experiments [30], there is a built-in stress to the stretched membrane over the hole. Thus, it is imperative to choose a peak that is relatively insensitive to the stretching of the crystal but with high enough χ_T. For most materials the thermal conductivity measurement can be performed with a 10 K average temperature increase. As an example χ_T for MoS_2 is about $-0.013\,cm^{-1}/K$. A very long focal length spectrometer equipped with a cutting-edge spectrometer, the resolution is limited to $0.5\,cm^{-1}$. Lorentzian fitting to the Raman peak can improve the peak position detection sensitivity to $0.25\,cm^{-1}$. Still, such a sensitivity requires 20 K of average temperature increase over the hole. Such a temperature rise might be very important in the study of materials that show temperature dependent phase transitions.

Absorbance of the laser power by the sample is another point that requires attention. As the calculation of the thermal conductivity depends on the laser power absorbed by the material, absorbance of the material has to be determined. Moreover, in most cases the hole under the sample is not a through hole and the light scattered from its base also contributes to the heating of the sample. Other limitations might also arise from the experimental setup. As an instance, positioning of the laser spot at the center of the suspended part is important. Any inaccuracy in the positioning of the laser spot will lead to large errors in thermal conductivity value for small radius holes. Another experimentally important aspect of the measurement is the beam profiling. Gaussian laser spot has to be profiled to extract the full-width-half-maximum.

2.4.2 Micro-Bridge Thermometry

Micro-bridge thermometry is an absolute thermal conductivity measurement method that relies on thermal isolation of the sample from the surrounding and delivering a measurable amount of heat from one side of the sample to the other while measuring the temperature of the sample at both sides. SEM images of the experimental configuration and the equivalent thermal circuit is given in Fig. 2.3. The method has been applied on measuring the thermal properties of multi-wall carbon nanotubes by Kim et al. [31]. The method has been then adapted for 2D materials. The method has been applied on graphene [5, 32–35], h-BN [36], MoS_2 [15] and black phosphorus [37].

2.4.2.1 Limitations of the Micro-Bridge Thermometry

Despite the versatility of the micro-bridge thermometry, implementation of the method is very difficult especially on 2D materials. Most of the studies cited for the micro-bridge thermometry uses either thick crystals or they require very stiff materials like graphene and h-BN to survive the intense device fabrication procedures.

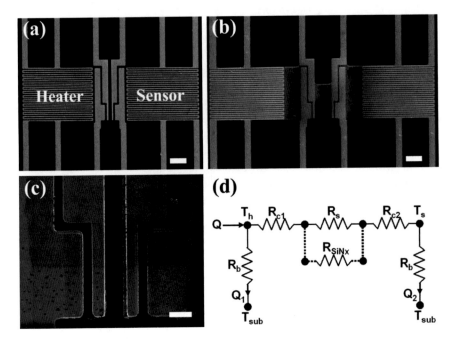

Fig. 2.3 Figure copied from Wang et al. [32] showing SEM micrographs of **a**. the micro-bridge structures with heater and the sensor parts, **b**. with the graphene flake suspended across the the pads and **c**. a close-up image of the contact pads. **d**. Equivalent thermal circuit of the device is shown. Scale bars are 5 μm for a and b, 1 μm for c. Reprinted with permission from Wang et al. [32]. Copyright (2011) American Chemical Society

Device fabrication steps including many lithography and etching also leads to significant contamination of the crystals, resulting in significant changes in the thermal properties due to the scattering of the phonons from the polymeric residues [32, 33]. Thus, the widespread use of the method is not possible. Moreover, similar to the Raman thermometry the stress on the crystal further complicates the analysis of the measurements. Perhaps, this is common to the all thermal conductivity measurement methods, yet for the sake of completeness we should also mention that it is not entirely possible to isolate the heating pads from the surrounding. As a result, the heat lost to the connections is very difficult to account for, leading to further uncertainties.

2.4.3 Scanning Thermal Microscopy

Scanning thermal microscopy (SThM) is a variant of AFM that is modified to measure the local temperature over the sample. Typically, the AFM tip is made of Pt-Cr junction to be used as a thermocouple that is overlayed on one another with separation terminated near the tip. The method first introduced by Majumdar and his group [38–40]. It is possible to measure the temperature distribution with a spatial resolution of 100 nm, that is more than twice better than the Raman thermometry. A quantitative measurement of the temperature, of course, requires a very careful analysis of all the possible heat loss mechanisms and interaction of the tip with substrate. Due to these complications it has not been widely used in the thermal conductivity measurements of 2D materials (Fig. 2.4).

2.4.4 Time-Domain Thermoreflectance Method

Time-domain thermoreflectance (TDTR) method uses the reflectance change of the material upon heating to measure the thermal conductivity via intense modelling. The method is a pump-probe spectroscopy in essence. An ultra-fast laser pulse sent on to the sample to create a local temperature rise due to optical absorption and then the second pulse, probes the reflection change of the sample. The time when the probe pulse with respect to the pulse arrives on the sample can provide how the heat is dissipated over the sample in time. The method is introduced by Paddock et al. [42] to measure the thermal diffusivity of metallic films. The laser spot is focused to a 20 μm in the original work and given the fact that the heat diffusion within a few hundred picoseconds can only reach a fraction of the laser spot, the model assumes a one-dimensional diffusion in to the sample. The heat diffusion length $l \approx \sqrt{h\tau}$ where h is the thermal diffusivity and τ is the time after the arrival of the heating pulse. For a typical metal with 200 ps delay between the pump and the probe, $l \approx 50$ nm. If the sample thickness is restricted below l then, effectively the problem is a 1D heat diffusion problem.

The method can be used to measure not only the thermal diffusivity of the sample but thermal properties of materials in general such as thermal conductivity [44, 45], the thermal conductance of interfaces [46, 47], microfabricated structures etc. [48]. To measure the thermal properties of 2D materials, a highly thermoreflective coating such as aluminium has to applied on semiconducting crystals. Besides having high thermoreflectance, Al has a linear dependence of reflactance to the surface temperature change. As 2D materials are highly anisotropic in in-plane and out-of-plane directions, one-dimensional model proposed above is not applicable to 2D materials. Further the interface between the Al layer and the 2D material has to be taken in to account.

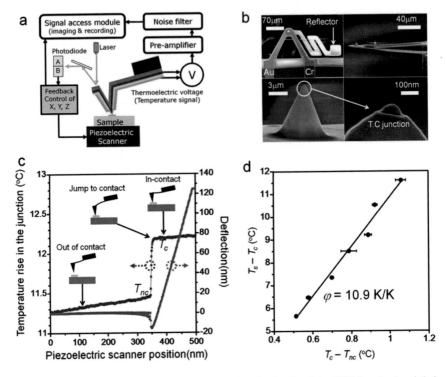

Fig. 2.4 Figure copied from Kim et al. [41] showing the details of the SThM method and their measurements. **a**. Schematic of the SThM operation. **b**. SEM micrographs of the AFM tip used by Kim et al. [41] **c**. Temperature rise at the junction with respect to the piezo position as well as the deflection of the tip versus the piezo position is given. Upon contact with the substrate, the temperature rises by almost a Kelvin. **d**. Difference between the sample temperature T_s and the contact temperature T_c versus the difference between T_c and the non-contact temperature T_{nc} gives the dimensionless constant ϕ for the particular tip. The figure is reprinted with permission from Kim et al. [41]. Copyright (2011) American Chemical Society

2.4.4.1 Limitations of the TDTR Method

The major limitation of TDTR method is that it cannot be really applied on 2D materials.[1] First of all, a minimum sample thickness of 20 nm is required for acquiring the thermal properties of the material. This comes at the cost of having cross-plane heat transport, which is highly different than the in-plane thermal transport in 2D materials due to the weak interlayer coupling. Jiang et al. [16] measured the in-plane and cross-plane thermal conductivities of MX_2 (M=Mo, W and X=S, Se) bulk crystals using TDTR technique using a variable-spot-size approach [49] and showed that there is nearly two orders of magnitude difference in cross-plane and in-plane

---

[1] Unless the purpose is to study the the cross-plane thermal conductivity, then TDTR is superior to all techniques especially with the recent improvements as discussed later.

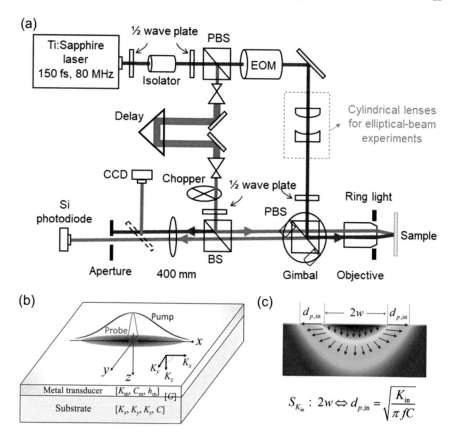

Fig. 2.5 a. Schematic showing the experimental setup for the TDTR method. EOM, PBS, and BS stand for electro-optic modulator, polarizing beam splitter, and beam splitter, respectively. **b.** Schematic showing how elliptical-beam method works on the sample. **c.** Heat flux directions and the comparison of the laser spot size to the sample depth. Figure is reprinted with the permission of Jiang et al. [43]. Copyright (2018) AIP Publishing

thermal conductivity values. The method is also applied on bulk black phosphorus [50], graphene and h-BN [49]. Finally, the very expensive and highly customized experimental setup makes the technique inaccessible to widespread use (Fig. 2.5).

2.4.5 3ω Method

The method requires a narrow metal line to be placed on the material to be measured, typically prepared by lithography followed by metal evaporation. This metal line serves both as the heater and the thermometer for the measurements. An ac current at 2ω is applied to the metal to cause Joule heating. This results in temperature oscillations of the metal line at frequency 2ω and as a result the resistance change of the metal can be detected at the third harmonic 3ω. To my best knowledge, the method has not been applied to 2D materials due to the insensitivity of method to the in-plane and cross-plane thermal conductivity. However, I wanted to include it among the measurement techniques as it is potentially applicable to atomically thin isotropic materials.

2.4.6 Bolometry Based Method

In a recent study, we introduced a novel method to measure thermal conductivity in 2D materials, in particular the metallic ones based on the bolometric effect [51]. In Chap. 3, I provide a very detailed description of the method with many details. Here, I wanted to introduce the method very briefly to give a background for the comparison table for the thermal conductivity methods that I provided in the next subsection. In essence the bolometry based thermal conductivity measurement method is very similar to the Raman thermometry based thermal conductivity measurement method.

2.4.7 Comparison of Thermal Conductivity Measurement Methods

I have to state that there is no single "best thermal conductivity measurement method" as all the techniques mentioned are superior to others in certain aspects. Thus, I would like to compare different thermal conductivity measurement methods not to find a better method but to demonstrate their domains of applicability. Moreover, it should be noted that a particular choice of the measurement method may extend beyond simply measuring the thermal conductivity of a material and to study various thermal properties of a material. For that matter, Raman thermometry has been the simplest choice for many researchers due to the ease of applicability to 2D materials thanks to very simple sample preparation that only requires the transfer of 2D flake over pre-drilled holes on the substrate. Although the experimental setup is rather complicated, micro-Raman setups are widely available in many research laboratories. One has to integrate a cooling or an heating stage as well. Moreover, a fine diffraction grating is required (typically something beyond 1200 lines/mm) to have a good spectral resolution. However, the method suffers from low accuracy and low sensitivity. Moreover, a moderate amount of data analysis is required to extract the thermal conductivity value. Method only works for in-plane measurements. In-plane anisotropy can be measured with a line profiled laser spot. TDTR method is a well established method that has been used very frequently in the study of thermal conductivity in many materials. However, its applicability to 2D materials is limited to cross-plane measurements. The greatest drawback of the method is the vast complexity of the

Method \ Attribute	Raman Thermometry	TDTR	Micro-Bridge Thermometry	Bolometry Based Thermometry
Applicability to 2D Materials	Applicable	Applicable in bulk	Limited Applicablility	Applicable
Ease of Sample Preparation	Easy	Moderate	Very Difficult	Moderate
Experimental Setup	Complicated	Very Complicated	Very Complicated	Complicated
Accuracy	Low	High	Very High	High
Sensitivity	Low	High	Very High	High
Data Analysis	Moderate	Intensive	Simple	Simple

Fig. 2.6 A qualitative comparison table to compare different thermal conductivity measurement methods. Details of the bolometry based thermometry is explained in Chap. 3

measurement setup. In essence it is a pump-probe spectroscopy that requires an ultra-fast laser with delay lines and many optical components. Also, it requires intensive data analysis to extract the thermal properties of the material. It provides high sensitivity and high accuracy compared to other methods. Micro-bridge thermometry method has rather limited applicability to 2D materials manly due to very difficult device fabrication steps. Moreover, the experimental setup for the measurement are relatively complicated. It provides very simple data analysis with very high accuracy and sensitivity. Details of the bolometry based thermometry is given in Chap. 3. To summarize, it is highly applicable to 2D materials that shows limited photo-voltaic photoresponse. Sample preparation is slightly more difficult than the Raman thermometry as it requires fabrication of metal contacts on the substrate. Experimental setup is fairly complicated as the measurement requires a current pre-amplifier attached to a lock-in amplifier. It provides high accuracy with very high sensitivity. Data analysis is also simple. Figure 2.6 gives a table for a comparison of the methods.

References

1. Sichel EK, Miller RE, Abrahams MS, Buiocchi CJ (1976) Phys Rev B 13(10):4607. https://link.aps.org/doi/10.1103/PhysRevB.13.4607
2. Schmidt AJ, Chen X, Chen G (2008) Rev Sci Instrum 79(11):114902. http://aip.scitation.org/doi/10.1063/1.3006335
3. Liu J, Choi GM, Cahill DG (2014) J Appl Phys 116(23):233107. http://dx.doi.org/10.1063/1.4904513, http://aip.scitation.org/doi/10.1063/1.4904513

4. Kim TY, Park CH, Marzari N (2016) Nano Lett 16(4):2439. https://doi.org/10.1021/acs. nanolett.5b05288
5. Seol JH, Jo I, Moore AL, Lindsay L, Aitken ZH, Pettes MT, Li X, Yao Z, Huang R, Broido D, Mingo N, Ruoff RS, Shi L (2010) Science 328(5975):213. https://www.sciencemag.org/ lookup/doi/10.1126/science.1184014
6. Lindsay L, Broido DA, Mingo N (2010) Phys Rev B 82(11):115427. https://link.aps.org/doi/ 10.1103/PhysRevB.82.115427
7. Lindsay L, Li W, Carrete J, Mingo N, Broido DA, Reinecke TL (2014) Phys Rev B 89(15):155426. https://link.aps.org/doi/10.1103/PhysRevB.89.155426
8. Wei Z, Yang J, Chen W, Bi K, Li D, Chen Y (2014) Appl Phys Lett 104(8):081903. http://aip. scitation.org/doi/10.1063/1.4866416
9. Fu Q, Yang J, Chen Y, Li D, Xu D (2015) Appl Phys Lett 106:3. http://dx.doi.org/10.1063/1. 4906348
10. Zhang H, Chen X, Jho YD, Minnich AJ (2016) Nano Lett 16(3):1643. https://pubs.acs.org/ doi/10.1021/acs.nanolett.5b04499
11. Chen S, Wu Q, Mishra C, Kang J, Zhang H, Cho K, Cai W, Balandin AA, Ruoff RS (2012) Nat Mater 11(3):203. http://dx.doi.org/10.1038/nmat3207, http://www.nature.com/articles/ nmat3207
12. Zou JH, Cao BY (2017) Appl Phys Lett 110(10):103106. https://doi.org/10.1063/1.4978434.
13. Sood A, Xiong F, Chen S, Cheaito R, Lian F, Asheghi M, Cui Y, Donadio D, Goodson KE, Pop E (2019) Nano Lett 19(4):2434. https://doi.org/10.1021/acs.nanolett.8b05174
14. Sahoo S, Gaur AP, Ahmadi M, Guinel MJ, Katiyar RS (2013) J Phys Chem C 117(17):9042. https://doi.org/10.1021/jp402509w
15. Jo I, Pettes MT, Ou E, Wu W, Shi L (2014) Appl Phys Lett 104:20. http://dx.doi.org/10.1063/ 1.4876965
16. Jiang P, Qian X, Gu X, Yang R (2017) Adv Mater 29(36):1. https://doi.org/10.1002/adma. 201701068
17. Muratore C, Varshney V, Gengler JJ, Hu JJ, Bultman JE, Smith TM, Shamberger PJ, Qiu B, Ruan X, Roy AK, Voevodin AA (2013) Appl Phys Lett 102(8):081604. http://aip.scitation.org/ doi/10.1063/1.4793203
18. Li X, Zhang J, Puretzky AA, Yoshimura A, Sang X, Cui Q, Li Y, Liang L, Ghosh AW, Zhao H, Unocic RR, Meunier V, Rouleau CM, Sumpter BG, Geohegan DB, Xiao K (2019) ACS Nano 13(2). https://pubs.acs.org/doi/10.1021/acsnano.8b09448
19. Cao Y, Fatemi V, Fang S, Watanabe K, Taniguchi T, Kaxiras E, Jarillo-Herrero P (2018) Nature 556(7699):43. https://doi.org/10.1038/nature26160
20. Gao Y, Liu Q, Xu B (2016) ACS Nano 10(5):5431. https://pubs.acs.org/doi/10.1021/acsnano. 6b01674
21. Zheng X, Zhao C, Gu X (2019) Int J Heat Mass Transf 143:118583. https://doi.org/ 10.1016/j.ijheatmasstransfer.2019.118583, https://linkinghub.elsevier.com/retrieve/pii/ S0017931019323312
22. Périchon S, Lysenko V, Remaki B, Barbier D, Champagnon B (1999) J Appl Phys 86(8):4700. https://doi.org/10.1063/1.371424
23. Périchon P, Lysenko V, Roussel P, Remaki B, Champagnon B, Barbier D, Pinard P (2000) Sens Actuators A Phys 85(1):335. https://doi.org/10.1016/S0924-4247(00)00327-7
24. Balandin AA, Ghosh S, Bao W, Calizo I, Teweldebrhan D, Miao F, Lau CN (2008) Nano Lett 8(3):902. https://doi.org/10.1021/nl0731872
25. Yan R, Simpson JR, Bertolazzi S, Brivio J, Watson M, Wu X, Kis A, Luo T, Hight Walker AR, Xing HG (2014) ACS Nano. 8(1):986. https://doi.org/10.1021/nn405826k
26. Zhang X, Sun D, Li Y, Lee GH, Cui X, Chenet D, You Y, Heinz TF, Hone JC (2015) ACS Appl Mater Interfaces 7(46):25923. https://doi.org/10.1021/acsami.5b08580
27. Wang R, Wang T, Zobeiri H, Yuan P, Deng C, Yue Y, Xu S, Wang X (2018) Nanoscale 10(48):23087. https://doi.org/10.1039/c8nr05641b
28. Peimyoo N, Shang J, Yang W, Wang Y, Cong C, Yu T (2015) Nano Res 8(4):1210. https://doi. org/10.1007/s12274-014-0602-0

29. Vallabhaneni AK, Singh D, Bao H, Murthy J, Ruan X (2016) Phys Rev B 93(12):1. https://doi.org/10.1103/PhysRevB.93.125432
30. Lee C, Wei X, Kysar JW, Hone J (2008) Science 321(5887):385. https://www.sciencemag.org/lookup/doi/10.1126/science.1157996
31. Kim P, Shi L, Majumdar A, McEuen PL (2001) Phys Rev Lett 87(21):215502. https://doi.org/10.1103/PhysRevLett.87.215502
32. Wang Z, Xie R, Bui CT, Liu D, Ni X, Li B, Thong JT (2011) Nano Lett 11(1):113. https://doi.org/10.1021/nl102923q
33. Pettes MT, Jo I, Yao Z, Shi L (2011) Nano Lett 11(3):1195. https://doi.org/10.1021/nl104156y
34. Jang W, Bao W, Jing L, Lau CN, Dames C (2013) Appl Phys Lett 103(13):6. https://doi.org/10.1063/1.4821941
35. Xu X, Pereira LF, Wang Y, Wu J, Zhang K, Zhao X, Bae S, Tinh Bui C, Xie R, Thong JT, Hong BH, Loh KP, Donadio D, Li B, Özyilmaz B (2014) Nat Commun 5:1. https://doi.org/10.1038/ncomms4689
36. Jo I, Pettes MT, Kim J, Watanabe K, Taniguchi T, Yao Z, Shi L (2013) Nano Lett 13(2):550. https://doi.org/10.1021/nl304060g
37. Lee S, Yang F, Suh J, Yang S, Lee Y, Li G, Choe HS, Suslu A, Chen Y, Ko C, Park J, Liu K, Li J, Hippalgaonkar K, Urban JJ, Tongay S, Wu J (2015) Nat Commun 6. https://doi.org/10.1038/ncomms9573
38. Majumdar A (1999) Annu Rev Mater Sci 29(1):505. http://www.annualreviews.org/doi/10.1146/annurev.matsci.29.1.505
39. Shi L, Plyasunov S, Bachtold A, McEuen PL, Majumdar A (2000) Appl Phys Lett 77(26):4295. http://aip.scitation.org/doi/10.1063/1.1334658
40. Shi L, Majumdar A (2002) J Heat Transf 124(2):329. https://doi.org/10.1115/1.1447939
41. Kim K, Chung J, Hwang G, Kwon O, Lee JS (2011) ACS Nano 5(11):8700. https://doi.org/10.1021/nn2026325
42. Paddock CA, Eesley GL (1986) J Appl Phys 60(1):285. https://doi.org/10.1063/1.337642
43. Jiang P, Qian X, Yang R (2018) Rev Sci Instrum 89(9):094902. http://aip.scitation.org/doi/10.1063/1.5029971
44. Cahill DG, Goodson K, Majumdar A (2002) J Heat Transf 124(2):223. https://doi.org/10.1115/1.1454111
45. Cahill DG, Ford WK, Goodson KE, Mahan GD, Majumdar A, Maris HJ, Merlin R, Phillpot SR (2003) J Appl Phys 93(2):793. https://doi.org/10.1063/1.1524305
46. Costescu RM, Wall MA, Cahill DG (2003) Phys Rev B Condens Matter Mater Phys 67(5):1. https://doi.org/10.1103/PhysRevB.67.054302
47. Costescu RM, Cahill DG, Fabreguette FH, Sechrist ZA, George SM (2004) Science 303(5660):989. https://www.sciencemag.org/lookup/doi/10.1126/science.1093711
48. Huxtable ST, Cahill DG, Phinney LM (2004) J Appl Phys 95(4):2102. https://doi.org/10.1063/1.1639146
49. Jiang P, Qian X, Yang R (2017) Rev Sci Instrum 88:7. http://dx.doi.org/10.1063/1.4991715
50. Jang H, Wood JD, Ryder CR, Hersam MC, Cahill DG (2015) Adv Mater 27(48):8017. https://doi.org/10.1002/adma.201503466
51. Cakiroglu O, Mehmood N, Çiçek MM, Aikebaier A, Rasouli HR, Durgun E, Kasirga ST (2016) 2D Mater 11(2020). https://iopscience.iop.org/article/10.1088/2053-1583/ab8048

Chapter 3
Thermal Conductivity Measurements via the Bolometric Effect

Abstract In this chapter I will introduce the measurement of thermal conductivity using the bolometric effect. The bolometric effect is defined as the resistivity change of a material due to heating. Indeed, the bolometric effect forms the basis of many modern technological sensors and devices. For instance, most commonly used integrated circuit thermometers are based on the well calibrated resistivity change of a Pt strip. Another example is the thermal imaging sensors. A cooled array of high temperature coefficient of resistance material can sensitively detect the infrared spectrum due to the change in the electrical resistance of the active material.

Keywords Bolometric thermal conductivity measurement method · Thermal conductivity in metallic 2D materials · Scanning photocurrent microscopy · Thermal conductivity measurements in 1D

3.1 Introduction

Almost all pure materials whether metallic or semiconducting show a well-defined temperature dependent resistivity in a wide range of temperature. Even if there is no temperature dependence that can be written as a well-defined continuous function, it is possible to describe the temperature dependence for certain temperature intervals. This temperature dependence can be used to determine the temperature distribution over a material even if the heat source is local and there is a temperature gradient over the material. The measurement process can be performed as the following. The electrical resistance of the sample is determined when it is in thermal equilibrium with the environment in the absence of the local heat source. Then, the local heat is applied to the sample and once the steady state is reached the electrical measurement is performed. The difference in the resistivities of the measurements before and after the heating is due to the temperature dependent resistivity change in the material. The thermal distribution can be modelled precisely with a known thermal conductivity. However, the purpose of the method is to measure the thermal conductivity. So, by using the thermal conductivity as a fitting parameter, using the thermal distribution, the resistance of the measured specimen can be calculated to match the experimental

© The Author(s), under exclusive license to Springer Nature Singapore Pte Ltd. 2020
T. S. Kasirga, *Thermal Conductivity Measurements in Atomically
Thin Materials and Devices*, Nanoscience and Nanotechnology,
https://doi.org/10.1007/978-981-15-5348-6_3

value. This strategy is not advantageous for measuring bulk materials as there are much more convenient and direct ways of measuring the thermal conductivity. However, in the case of a nanowire or a nanosheet, when the number of available methods are limited, the technique has been proven to be useful.

3.2 Theoretical Background

A focused laser beam is a very convenient way to deliver heat onto many 2D materials as in the case of Raman thermometry. The major concern here is the existence of several photoconductance mechanisms that limits the usability of the technique in semiconducting materials. However, no single technique in the realm of thermal conductivity measurements rule them all as discussed before. So far, by using the laser beam as a local heat source, we measured the thermal conductivity of metallic thin sheets. However, via a detailed analysis, the photoconductance mechanisms can be differentiated experimentally and the bolometric effect can still be used to extract the thermal conductivity.

When a 2D material is laid over a smooth substrate, a considerable portion of the heat will be transferred to substrate through the interface. Such a measurement is very challenging for a 2D material. Thus, ideally if the 2D sheet can be suspended across a trench or over a circular hole, the requirement to measure the thermal boundary conductance will be eliminated. This is discussed in the previous chapters and we showed that the Raman thermometry based method uses the same experimental methodology.

3.2.1 Bolometric Effect

When the electromagnetic radiation impinging on a material causes a change in the electrical resistivity of a material due to the raising temperature, the calibrated change of the electrical resistance of the device can be used as a thermometer. The effect has many applications ranging from astronomy to particle physics and micro-bolometer arrays are used commonly for thermal imaging.

The electrical resistance of most metals and semiconductors depend on the temperature around the room temperature. For the metals, the electrical resistance increases linearly with the increasing temperature. At cryogenic temperatures, typically below 20 K, residual resistivity appears and a plateau in the resistance-temperature curve is observed. For many metals, if the temperature is further decreased, the superconducting state appears where the electrical resistivity vanishes. At very high temperatures deviations from the linearity is observed due to the combined effects of increased scattering, vacancy formation and strain. The general expression for the electrical resistivity ρ can be written as:

$$\rho(T) = \rho_0[1 + \alpha(T - T_0)] \tag{3.1}$$

where T_0 is a fixed reference temperature, α is the temperature coefficient of resistance, ρ_0 is the resistivity at T_0. For the semiconductors, a similar expression can be written. In the case of semiconductors, the number of electrical carriers increase exponentially due to the change in the number of electrons that are be transferred to the conduction band. It is rather straightforward to derive the following relation for an intrinsic semiconductor:

$$\rho(T) = \rho_0 e^{E_A/k_B T} \tag{3.2}$$

Here, E_A is the activation energy and k_B is the Boltzmann constant. Activation energy is half of the band gap for some semiconductors however, it is better to be determined experimentally from the resistance versus temperature (RT) measurements. Details of RT measurements is given in the experimental section.

3.2.2 Analytical Solution of the Heat Equation for Isotropic 2D Materials

For the device geometry given in Fig. 3.1. It is straightforward to solve the heat equation over the suspended part of the material as well as the supported part with a Gaussian laser spot delivering the heat to the system. For the sake of simplicity we excluded the Newtonian cooling. Heat equation for the suspended part of the crystal

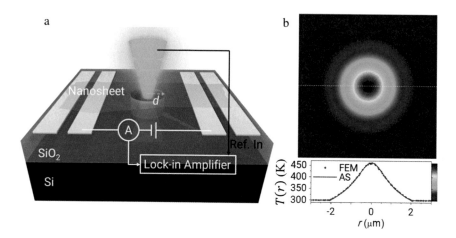

Fig. 3.1 **a** Schematic of the device and the measurement scheme is given. The crystal is suspended over a hole of radius d. **b** Temperature distribution over and in the vicinity of the hole for a typical 2D material under typical laser illumination power. Lower panel shows the line trace taken along the center of the suspended section showing the analytical solution and the FEM solution

in radial coordinates to find the temperature distribution over the hole, $T_1(r)$, can be written as:

$$\kappa \frac{1}{r}\frac{d}{dr}\left[r\frac{d}{dr}T_1(r)\right] + q(r) = 0 \qquad r < R \qquad (3.3)$$

where

$$q(r) = \frac{P\alpha}{\pi t r_0^2}e^{-r^2/r_0^2} \qquad (3.4)$$

is the heat delivered by the volumetric Gaussian laser spot at the center of the hole. P is the laser power, r_0 is the full width at half maximum of the Gaussian spot, α is the absorbance of the crystal, t is the thickness of the crystal. Outside of the suspended part, where $r > R$, the heat equation is written as the following:

$$\kappa' \frac{1}{r}\frac{d}{dr}\left[r\frac{d}{dr}T_2(r)\right] - G/t[T_2(r) - T_0] = 0 \qquad r > R \qquad (3.5)$$

Here, a different value to the thermal conductivity κ' is assigned to the supported part of the material due to the substrate effects on the 2D materials as discussed in Chap. 2. $T_2(r)$ is the temperature distribution over the supported part and T_0 is the ambient temperature. G is the thermal boundary conductance between the crystal and the substrate. Solving Eqs. 3.3 and 3.5 together with the appropriate boundary conditions yields the temperature distribution everywhere on the sample. The general solutions to the above equations give:

$$T_1(r) = c_1 + c_2 ln(r) + \frac{\alpha P}{4\pi t\kappa}Ei(-\frac{r^2}{r_0^2}) \qquad r < R \qquad (3.6)$$

$$T_2(\gamma) = c_3 I_0(\gamma) + c_4 K_0(\gamma) + T_0 \qquad r > R \qquad (3.7)$$

where $\gamma = r\left[\frac{G}{\kappa' t}\right]^{1/2}$, introduced for the ease of calculations. I_0 and K_0 are the zeroth order modified Bessel's functions of the first and the second kind, respectively. Ei is the exponential integral. As for the boundary conditions we use the following to find the constants c_1, c_2, c_3 and c_4:

$$(i) \qquad T_1(R) = T_2(R) \qquad (3.8)$$

$$(ii) \qquad T_2(\gamma \to \infty) = T_0 \qquad (3.9)$$

$$(iii) \qquad \frac{dT_1(r)}{dr}\bigg|_{r\to 0} = 0 \qquad (3.10)$$

$$(iv) \qquad \kappa\frac{dT_1(r)}{dr}\bigg|_{r\to R} = \kappa'\frac{dT_2(\gamma)}{dr}\bigg|_{r\to R} \qquad (3.11)$$

Under these boundary conditions we find the coefficients as the following:

$$c_1 = T_0 + \frac{\alpha P K_0(\gamma_R)}{2\pi Rt K_1(\gamma_R)}\sqrt{\frac{t}{G\kappa'}}\left[1 - e^{-\frac{R^2}{r_0^2}}\right] + \tag{3.12}$$

$$\frac{\alpha P}{4\pi t\kappa}\left[2\ln(R) - \text{Ei}\left(-\frac{R^2}{r_0^2}\right)\right]$$

$$c_2 = -\frac{a r_0^2}{2\kappa} \tag{3.13}$$

$$c_3 = 0 \tag{3.14}$$

$$c_4 = \frac{a r_0^2}{2R K_1(\gamma_R)}\sqrt{\frac{t}{G\kappa'}}\left[1 - e^{-\frac{R^2}{r_0^2}}\right] \tag{3.15}$$

Under these constant, using Eqs. 3.6 and 3.7 we can find the radial thermal distribution. Figure 3.1 shows the temperature distribution map for a typical material illuminated under 70 μW.

3.2.3 Calculating the Thermal Conductivity via the Bolometric Effect

The key element in the bolometric thermal conductivity measurement method is the calculation of the temperature dependent resistance change over the sample. As mentioned earlier, every material shows temperature dependent electrical resistance change to a certain extent. For metals, in the vicinity of the room temperature, the change is linear in temperature and very predictable. A resistivity-temperature measurement around the room temperature gives how overall resistivity of the material changes. This change can be correlated with the change over the suspended part of the material for the bolometric effect based thermal conductivity measurement method. Here, the aim is to fit the thermal conductivity value to the set of equations we will derive in the following part to measured resistivity change over the center of the hole (Fig. 3.2).

The challenge is calculating the total resistance change of the device when the laser is at the center of the suspended part of the crystal. If we assume that the center of the hole is at $\mathbf{r} = 0$, then the resistivity distribution can be written as $\rho(\mathbf{r}; T(\mathbf{r}))$. For a resistivity distribution over a 2D material the electrical current flow has to be solved in between the contacts. For samples with small temperature dependent coefficient of resistance (TCR), the equipotential lines can be assumed to be parallel to the contacts. Thus, by calculating the total resistance of the device by series and parallel addition of infinitesimally small resistive regions and stripes formed by these regions gives the total change of resistance. Although it is straightforward, it requires a numerical solution. However, for large TCR samples or measurements with large temperature gradients (i.e. high laser power) the equipotential lines become warped around the suspended part. Thus, the solution offered above is not applicable anymore and the local electric field, $\mathbf{E}(\mathbf{r})$, has to be solved along with the continuity equation for

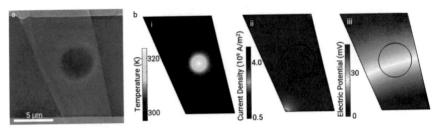

Fig. 3.2 **a** SEM micrograph of a measured 2H-TaS$_2$ crystal. **b** Thermal simulation of the same crystal, (**b**). Current density under 50 mV bias and **c**. The electric potential distribution over the sample. Notice the slight curvature of the equipotential points over the suspended section of the crystal. Due to increased local resistivity, potential drop is faster. Figure is reproduced from Cakiroglu et al. [1]. Copyright (2020) IOP Publishing

the current density, $\mathbf{J}(\mathbf{r})$. A numerical solution can be obtained for the resistance of the device. However, commercially available finite element method (FEM) solution packages provide a much more convenient way to solve the above problem. In the next section, I will talk about implementation of the problem to the commercially available FEM packages, in particular COMSOL®.

3.3 COMSOL Simulations for Thermal Conductivity Calculation

COMSOL Multiphysics uses finite element analysis to solve coupled differential equations in a wide range of physical problems. For our purposes, we use the Multiphysics and the heat transfer module. We also define the Gaussian laser spot and the heat flux from the laser using equation based solver. The workflow is as the following:

1. Measure the device dimensions. This requires measurement of the sample thickness in a suitable equipment, most typically using atomic force microscopy (AFM). Optical images can be used to measure the lateral size of the crystal.
2. Draw the device geometry in COMSOL with the suspended part defined as a material with different thermal conductivity.
3. Define the heat transfer parameters:

   - Define thermal boundaries
   - Define thermal regions
   - Define the heat in-flux over the suspended circular section.

4. Define the electrical terminals and the electrically isolated edges.
5. Enter the experimentally determined parameters and run the simulation to match the experimentally determined resistance change when the laser spot is at the center of the hole.

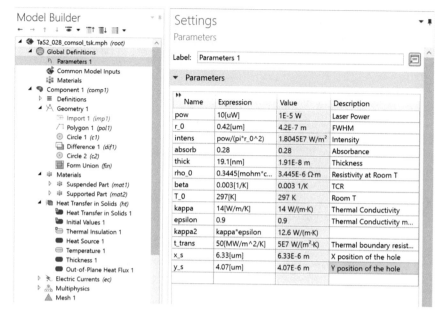

Fig. 3.3 Screenshots from the COMSOL Multiphysics software. Left panel shows the model parameters that are used to create the desired geometry and material parameters for the simulation. There is a single component in the simulation and all the geometries and relevant thermal and electrical boundary conditions are defined over the material. Right panel shows the parameters that are used as the inputs in for the simulation. Here, all the parameters are measured experimentally except the thermal conductivity and thermal conductivity over the supported section of the material. A parametric sweep can be defined in COMSOL to match the experimental δR value

Figure 3.3 shows the model builder and the parameter settings screenshots of the COMSOL graphical user interface. Figure 3.4 shows the model generated for the simulations as well as the multiphysics coupling between the electrical and the thermal simulations for the resistance calculations.

3.3.1 Uniqueness of the Thermal Conductivity Values

One question that begs for an answer is the uniqueness of the obtained thermal conductivity values. We calculated that for a set of κ values, the thermal equation solutions yield a unique δR value. Having a unique value for the resistance change caused by the laser heating is very important for the reliability of the measurement method. Figure 3.5a shows the results of the COMSOL simulations for a set of thermal conductivity values. Another important parameter as discussed in the Sect. 3.4.4 is the absorbance. Figure 3.5b shows the effect of absorbance on the simulated δR value and the fitted thermal conductivity. Moreover, we plotted the thermal profile over the cross-section of the suspended region to show each thermal distribution results in a unique resistivity distribution.

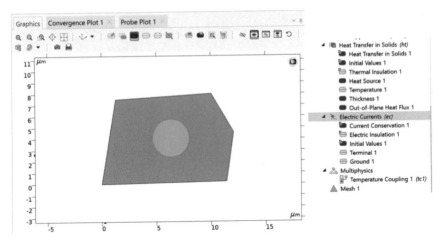

Fig. 3.4 Left panel shows the geometry of the sample defined in COMSOL. Green part at the center is the suspended region of the crystal whilst other regions are supported by a substrate. There is heat loss defined to the substrate in these regions. The panel on the right shows the electrical simulation parameters as well as the temperature coupling multiphysics package that relates the local temperature to the local resistivity of the material and solves for the total resistance of the material under illumination

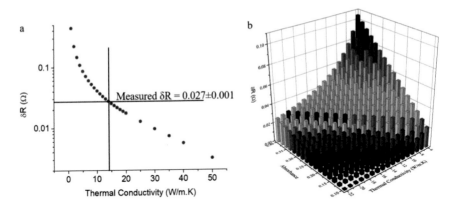

Fig. 3.5 **a** Graph shows how the calculated δR value varies for various thermal conductivity values. It is clear from the solution that there is a unique solution for each κ value. **b** When we vary two parameter at the same time, then we can estimate the propagation of error in the measured quantity to thermal conductivity fitting. Here absorbance is varied. Only the black bars are within the error margin of the measured δR value. Measurement error of the absorbance is less than ± 0.03 and this introduces a ± 1.0 W/m K error in the thermal conductivity value within the uncertainty of the δR measurements

3.4 Experimental Realization of the Thermal Conductivity Measurements via Bolometric Effect

In this Sect. 3.1 will talk about the experimental methods we use to perform the thermal conductivity measurements via the bolometric effect on 2D materials. These experimental methods include the sample preparation steps for the measurements as well as the experimental setup required for the high sensitivity measurements. First, the fabrication steps for the substrates that are required for the bolometric measurements is going to be discussed. Then, the commonly used transfer methods is going to be introduced. Finally, I am going to introduce the details of the experimental setup for the measurements.

3.4.1 Substrate Preparation

Although it may seem trivial, substrate preparation is a critical step of the bolometric thermal conductivity measurement method. The holes of radius ranging from 0.5 to 1.5 µm has to be drilled over the substrate with high aspect ratio walls. Furthermore, for the electrical resistance change measurements electrical contacts are required. Here, I will discuss various methods to fabricate the substrates for the measurements. The methods vary only with respect to the method the hole is etched on the substrate:

Method 1—With Focused Ion Beam Milling

- Chromium evaporation
- Photoresist spin coating
- Optical pattern forming
- Chromium etch
- Inductively coupled plasma etching by the depth of the metal to be deposited—This step is optional for easier transfer of the crystals on the metal contacts
- Metal deposition and lift-off
- Chromium etch
- Focused Ion Beam drilling of the hole in between the contacts
- Crystal transfer over the contacts and the hole.

Method 2—Without Focused ion Beam Milling

- Chromium evaporation
- Photoresist spin coating
- Optical pattern forming
- Chromium etch
- Inductively coupled plasma etching by the depth of the metal to be deposited—This step is optional for easier transfer of the crystals on the metal contacts
- Metal deposition and lift-off
- Photoresist spin coating

- Optical pattern forming for the holes—aligned with the substrate
- Chromium etch
- Etching of the holes—buffered oxide etch or inductively coupled plasma etch
- Resist removal and chromium etch
- Crystal transfer over the contacts and the hole.

The requirement for the electrical contact deposition introduces new steps as compared to the substrate preparation for the Raman thermometry based measurement technique. However, this is not a major challenge in device fabrication as the steps introduced for the electrical contact formation is a routine for many laboratories working on relevant materials.

The methods discussed above will have different parameters optimized for different systems. Here, the recipes I share are optimized for the equipment located at Bilkent University National Nanotechnology Research Center.

3.4.1.1 Chromium Evaporation

Chromium evaporation is required for forming an hard mask for the etching steps as most of the optical resists get badly damaged during the ICP etching. Si wafer with 1000 nm SiO_2 is cleaned thoroughly using acetone, IPA and water. Then, we evaporate 20 nm thick chromium using electron beam evaporator at a rate of 5 Å per min.

3.4.1.2 Optical Lithography

Once the Cr is deposited on the chips, AZ5209 photoresist is spin coated on and pre-baked before UV patterning. AZ5209 is a negative tone resist. The use of negative tone resist is crucial as in the liftoff process they prevent sidewall formation due to undercutting as compared to the positive tone resists. I will refrain from providing a recipe for the optical lithograph process as it is a very generic process. The thickness of the resist is important as the following step requires ICP etching and it is important to have sufficient amount of photoresist left after the etching process. Following the development of the exposed parts, Cr etching is used to expose the surface of the SiO_2.

3.4.1.3 ICP Etching

Inductively coupled plasma etching steps uses CH_3F and O_2. The coil and the plate are operated at 13.56 MHz and 396 KHz at 250 W and 30 W respectively. The process worked at room temperature for 12 min. This etches a pit of depth of 65 nm on SiO_2 surface. This process is important for the metal deposition step to be followed next as explained in detail. I would like to note that BOE etching would also work for a certain extent as the undercut of the etching will be less than 1% of the width of the unetched gap.

3.4.1.4 Metal Deposition

Following the ICP etching we fill the open holes with Au/Cr. First, 20 nm of Cr is evaporated as an adhesion and filling layer, then 45 nm of Au is evaporated using thermal evaporator. Here, tooling factors of the thickness monitors have to be very accurately determined. Otherwise, there is a great chance of having an under filled or an overfilled hole. Within ±10 nm, metal deposition provides a surface sufficiently flat enough for the crystal transfer. When the depth is not accurately matched, the PDMS stamping transfer fails due to the poor contact of crystal with the substrate. Moreover, it becomes more likely to trap air between the hole and the crystal as well as the edges of the metal contacts and the crystal. Thus, having the correct amount of metal is important for high yield device fabrication and high accuracy thermal conductivity measurements.

3.4.1.5 Focused Ion Beam Milling

A hole between the electrical contacts are required for creating a suspended section of the crystal. For this reason, we used focused ion beam milling. FIB provides a fast and very high aspect ratio way to drill the hole of the desired size. Some examples of the FIB drilled holes in between the metal contact are shown in Fig. 3.6. The major shortcoming of the method is the relative slow speed of the number of holes that can be drilled on a multi-sample chip and FIB is a costly equipment to access and operate.

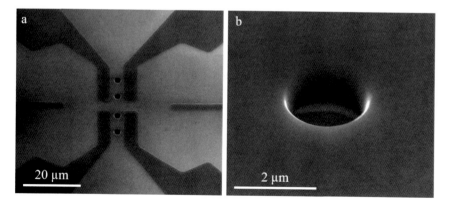

Fig. 3.6 **a** FIB drilled holes between the metal contacts and **b** Close up view of the hole is shown

3.4.1.6   Crystal Transfer

Crystal transfer is an important step. Placing the crystal on the gold contacts with minimal contact resistance is crucial. This has been demonstrated in an earlier study by our group [2]. When the contact resistance dominates the overall resistance of the device, the photoresponse becomes very localised at the contacts and the bolometric response diminishes. Another major challenge is that transferring the crystal over the hole without damaging the suspended section of the crystal. This mandates stamping of the crystal rather than polymer transfer method. A polydimethyl siloxane (PDMS) stamp is prepared using 1:7 ratio mix of the curing agent and the elastomer base. Once the mixture is thoroughly mixed, air bubbles trapped in the mixture are evacuated using a vacuum desiccator. Then the mixture is poured over a glass petri dish cleaned with oxygen plasma. This step is very important for the sake of consistency of the PDMS stamp. Using a PET petri-dish or skipping the plasma treatment results in variation in the stamp properties from batch to batch. This variation determines the overall efficiency of the transfer. With a properly prepared PDMS stamp the yield can as high as 90%, i.e. 9 crystals get successfully adhered to the substrate out of 10 trials. Some research groups use Gel-Pak brand in order to have a reproducible elastic stamp [3]. Mechanical exfoliation of the crystal from the bulk is performed using Nitto brand SPV224 blue tape from the bulk crystals commercially available from hqgraphene.com or 2dmaterials.com. Once, the crystals thinned down to the desired thickness, exfoliation is performed on to the PDMS surface. Thin crystals are identified under an optical microscope. Then the stamp is transferred over a glass slide and the glass slide is attached to a stepper motor driven 3-axis micromanipulator. The target substrate prepared as described in the previous parts is placed on to a heated stage. The crystal is aligned so that the contacts and the hole is centered. Then, by slowly bringing the substrate and the stamp in contact the transfer is performed. Slight heating up to 50 °C is needed for certain crystals to increase the adhesion to the substrate.

To summarize, I would like to emphasize the following points one more time:

- Thin metal contacts, preferably embedded contacts will improve the adhesion of the crystals.
- Stamping of exfoliated crystals provide a break free way of placing the them over the hole
- The whole process can be performed under 10 min from exfoliation to measurement. This minimizes the contact resistance formation as well as protects the device surface from oxidation.

3.4.2   Scanning Photocurrent Microscopy

Scanning photocurrent microscopy (SPCM) is a useful tool to provide the spatial distribution of the photoresponse of a material. A focused laser spot, raster scanned over a sample with metal contacts allows the creation of a photocurrent map. The setup is typically composed of a laser beam modulated at a specific frequency f. This frequency is fed to a lock-in amplifier and the output of a current preamplifier is attached to the input of the lock-in amplifier to measure the signal induced by the laser beam. This ensures that only the laser-induced photocurrent is measured. For feasible enough scanning speeds, typical integration time of 3 ms is chosen. Thus, any photoresponse mechanism that leads to photocurrent generation faster than 3 ms can be measured. Various wavelengths can be employed in SPCM measurements. Of course, depending on the wavelength of the laser used, the minimum feature size that can be resolved varies.

Laser scanning in a typical SPCM setup, the laser beam is focused on to the sample using a high resolving power objective. Typically we use a 40× objective. This creates a spot size of 400 nm FWHM for a 532 nm laser wavelength. The scan can be performed in three different ways. First, a galvo-mirror pair can be used to steer the light over the entrance pupil of the objective and this can move the focused spot on the sample. The range of the motion is limited to the size of the entrance pupil. Second scanning method can be scanning the objective. This is similar to the galvo-scanning, however in this method, the objective is attached to a scanner and the steering of the laser beam is performed by the motion of the objective with respect to the fixed laser beam. The third method would be scanning the sample with all the optics held fixed. This method increases the stability of the optical path at a cost of reduced reduced scan range. The setup we use is a commercially available setup from a start-up company where they compacted all the optics so that the whole scanning is performed by a moving microscope. This allows a very large range of motion, which becomes very handy when there are multiple samples on a chip inside the cryostat. The schematic and the picture of the setup is shown in Fig. 3.7.

Lock-in amplification for a sensitive measurement of the resistance change, we use the lock-in amplification scheme. Lock-in amplifier uses the orthogonality of a single frequency component to other harmonics in a noisy signal to extract the signal to be measured even the noise is about 1000 times larger than signal to be measured. Since lock-in method is an ac measurement technique, any slow varying or dc component of the signal is also rejected. Thus it is possible to measure the effects only induced by the laser.

The photocurrent is measured from two terminals on the sample. The voltage bias can be applied from one of the contacts and the other contact acts as a virtual ground, connected to a current pre-amplifier. We use SR-560 as the current pre-amplifier. Although it is not mandatory it serves two purposes. First, it is a very low impedance equipment and can operate in a very large range device resistances, from a few ohms to hundreds of megaohms. Second, there are configurable filters that helps rejection of the signal that is out of the frequency band of interest. Another way to use SR-560 is to use it as a voltage bias source and have a real ground on the other terminal. It provides a voltage output resolution of 0.5 mV, which is adequate for many measurement scenarios. SR-560 outputs the amplified signal from −5.4 to 5.4 V depending on the sign of the measured current. This signal is fed to the

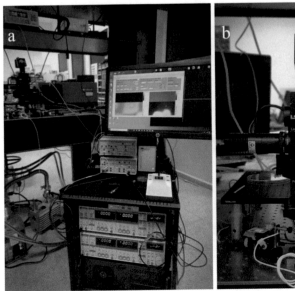

Fig. 3.7 a Photograph of the control box for the SPCM setup. **b** Close-up shot of the SPCM scanning head. The copper stage at the base is a Peltier stage with heating and cooling. Electrical probes are used for the photocurrent measurements

lock-in amplifier's voltage input. The laser beam is typically chopped around a kHz. This is slow enough for many processes, but sufficiently fast to reject any common noise sources like the line noise or its higher harmonics. The chopping can be both done electronically, with TTL output from the lock-in amplifier or with a mechanical chopper in the optical path, whose frequency is fed to the lock-in amplifier as the reference signal. Thus, the lock-in amplifies the signal that is resulting from the referenced laser beam. This, combined with the scanning motion allows the creation of the spatial photocurrent map. Here, the scanning has to be performed slower than the integration time of the lock-in. Typically we scan with a rate of 10 ms per pixel. Thus, we limit the integration time to maximum of 3–10 ms. Anything faster than 3 ms results in large noise for small currents.

Creating an intensity map from the reflected light to be able to correlate the photocurrent map to the parts of the sample, an intensity map from the reflected light is mandatory. This is done by placing a photodiode in the reflected light path. Since the laser beam is chopped, it is necessary to measure the signal with a lock-in amplifier as well. If cost of the setup is a concern, a phase locked loop that can be purchased in the form of audio amplifier serves the same purpose as the signal output of the photodiode is already significant enough. Another method would be tuning the frequency of the chopping and the scan rate so that there is enough intensity in every pixel. However, this would require a very stable mechanical chopper or a

TTL source. The reflection map also serves to determine the beam size using the knife-edge method. Spatial derivative of the intensity line trace from the edge of a beyond resolution limit metal contact provides a method to determine the FWHM of the laser spot.

Collection of the data we use LabView from National Instruments for the collection of the data since the ecosystem contains all the require drivers for instrument communication and the learning barrier for the programming with graphical user interface is very low. Furthermore, modifications to the software is very straightforward and can be done on the fly. However, Matlab or any other coding language can be used for the collection of the data. The software should do the following tasks:

- Control the scan in x-y direction with control over the scan size and the step size
- Control the experimental parameters such as the applied bias, instrument sensitivities etc.
- Collect and display the data in real time. This is very important as each fine scan takes around several minutes.
- Record and -preferably- analyse the data.

The intensity map and the corresponding photocurrent map provides a diffraction limited resolution information about the light induced electrical current. Thus the origin of the photoresponse can be identified by analysing the data.

3.4.3 Determining the Full Width-Half Maximum Value for the Focused Laser Spot

There are various tools for determining the profile of a laser beam as simple as a CMOS camera. However, I would like to introduce a method that is commonly used to determine the full width at half maximum (FWHM) of a Gaussian laser beam. As our method significantly benefits from scanning of the sample due to increased number of data points from the center of the suspended region, corresponding reflection map is produced for spatial confirmation of the position of the hole. Gold contacts with a large aspect ratio, allows a knife-edge determination of the laser profile. The position derivative of the line trace taken along the edge of the gold contact in the intensity map gives the cross-section of the laser beam. Gaussian fitting to the $\frac{\partial I}{\partial x}$ versus x plot yields the FWHM value for the laser spot. This might seem like a trivial aspect of the experimental scheme we proposed, however, on the contrary it significantly affects the uncertainty in the determination of the thermal conductivity value as r_0 is an input parameter in both Eqs. 3.3 and 3.4. Figure 3.8 shows the knife-edge determination of the laser FWHM.

In our experimental setup we use a 40x ultra-long working distance objective with slip-cover compensation up to 2 mm from Olympus. This objective pointed through a 1 mm sapphire window yields 430 nm FWHM for 532 nm laser. This is quite acceptable for an objective with 0.6 NA. Further improvement is possible if a higher magnification objective with larger numerical aperture is used.

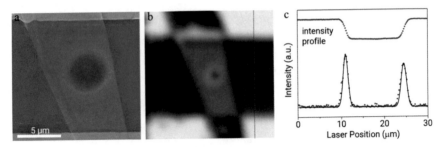

Fig. 3.8 **a** SEM micrograph of a 2H-TaS$_2$ crystal over the contacts. **b** SPCM reflection intensity map of the same crystal. Red dashed line shows the line trace given in **c** First derivative of the reflection intensity with respect to the position gives two Gaussian peaks. Gaussian fits to the peaks give the full width at half maximum value for the laser spot

3.4.4 Determining the Absorbance of the Material

Perhaps one of the most challenging aspects of the bolometric method is the measurement of the absorbance, α, of the material under investigation. For many materials, in particular for 2D materials, α is unknown. Moreover, it strongly depends on the number of layers when crystals are thinned to the monolayer limit. To measure the absorbance, we use the following method: First exfoliate the crystal on a transparent substrate. PDMS is a better choice as the thin flake identified on the PDMS stamp can later be transferred over the device for the thermal conductivity measurements. The laser intensity, I_0 is measured using a powermeter or a CCD camera attached to a spectrometer. Then, the transmitted intensity, T_0, is measured over the PDMS surface next to the specified crystal followed by the measurement of the transmitted intensity over the crystal, T. Finally, the reflected intensities from the bare PDMS surface, R_0, and the crystal surface, R, are measured. A schematic of the setup is shown in Fig. 3.9. Absorbance can be calculated from the quantities as the following:

$$\alpha = \frac{I_0 - (T + R) - (T_0 + R_0)}{I_0} \tag{3.16}$$

This measurement ensures that the correct ratio of the laser power absorbed can be used in the fitting of the experimental data. Even though for many materials α is temperature independent for a large range of temperatures, one has to be careful in particular when working with materials that show phase transitions which can significantly alter the absorbed laser power.

In our recent study, to measure the absorbance of 2H-TaS$_2$ for the first time we used a setup similar to that is described in the previous paragraph. The optical connection to the spectrometer is made through a multimode optical fiber with 200 μm core diameter. The end of the fiber is placed in contact with the 0.5 mm PDMS piece. The crystal is centered with the center of the fiber core to ensure maximum amount of light coupling to the fiber. The other end of the fiber is connected to an Andor

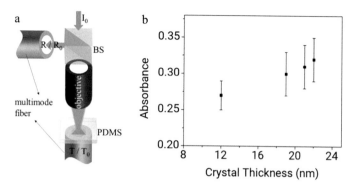

Fig. 3.9 a Schematic of the absorbance measurements is given in the figure. Both the transmitted light and the reflected light are measured using a spectrometer. **b** Thickness dependent absorbance of 2H-TaS$_2$ crystals. Data from

Shamrock 500i spectrometer equipped with iDus 400 CCD camera. We used 150 lines/mm diffraction grating to collect the intensity. As the optical losses within the spectrometer is the same for all measurements this is not too critical, however having the best possible signal to noise ratio decreases the measurement errors.

The effect of α on the thermal conductivity measurements is limited. COMSOL simulation in Fig. 3.5 shows how error in α propagates to the measured thermal conductivity value.

3.5 Extending the Bolometry Based Thermal Conductivity Measurements to 3D

It is possible to extend the methodology of the bolometry based thermal conductivity measurement method to take measurement in three-dimensional samples. The assumption we make in 2D bolometric measurements is that laser power is absorbed more or less uniformly through the thickness of the material and the absorbed heat from this point spreads uniformly through the sample in-plane axis. When the sample becomes thick enough, it is possible to model the thermal distribution in 3D. For thermally isotropic materials modelling the temperature distribution, thus the resistivity distribution is straightforward with the known optical absorbance of the material. Thermal conductivity can be measured using the method. For the bulk of 2D materials, anisotropy of the thermal conductivity is very large and this would require simultaneous fitting of both in-plane and cross-plane thermal conductivity values to match the experimentally measured resistance change. Thus, the uniqueness of the resistance solution will be questionable as there will be a set of thermal conductivity values that may give the same resistance change under the illumination. One solution to this problem could be measuring the in-plane thermal conductivity of a thin enough for uniform depth heating (but thick enough to be considered as bulk), then using the this value to find the cross-plane thermal conductivity in thicker samples. More experiments have to be conducted to fully understand the limitations of the bolometric method extended to 3D.

3.6 Extending the Bolometry Based Thermal Conductivity Measurements to 1D

The solutions of the heat equation becomes much simpler in 1-dimensional heat transport. In an earlier work [4], we used the thermal conductivity of VO_2 nanowires to calculate the bolometric resistance change. At the time the photoresponse mechanism for VO_2 was unknown, however, revealing the bolometric nature of the photoresponse shows us that whenever the photoresponse mechanism is known, it is possible to use the that to extract the thermal conductivity. 1D heat transport solutions can also be used in 2D materials with in-plane thermal conductivity anisotropy such as black phosphorus [5] when combined with clever measurement strategies. Rather than a Gaussian laser spot, a line shaped illumination along the crystal width can be used to extract the thermal conductivity along a certain axis of the crystal.

1D heat equation ignoring the Newtonian cooling and heat loss by radiation can be written as:

$$\kappa \frac{d^2 T_1(x)}{dx^2} + \frac{I\alpha}{t} e^{-x^2/w_0^2} = 0 \qquad x < w \qquad (3.17)$$

$$\kappa' \frac{d^2 T_2(x)}{dx^2} - \frac{G}{t}(T_2(x) - T_0) = 0 \qquad x > w \qquad (3.18)$$

Here, w is the width of the suspended part of the crystal. All other parameters are defined in Sect. 3.2.2. General solutions to the above equations are:

$$T_1(x) = C_1 + C_2 x - \frac{a\sqrt{\pi}w_0}{2\kappa}\left(\frac{w_0}{\sqrt{\pi}}e^{-x^2/w_0^2} + erf(\frac{x}{w_0})x\right) \qquad (3.19)$$

$$T_2(x) = T_0 + C_3 e^{x\sqrt{\frac{G}{t\kappa'}}} + C_4 e^{-x\sqrt{\frac{G}{t\kappa'}}} \qquad (3.20)$$

where erf is the error function and C_i are constants of integration. When we introduce the following boundary conditions, we can write the solutions of the differential equation:

$$(i) \qquad T_2(x \to \infty) = T_0 \qquad (3.21)$$

$$(ii) \qquad \frac{dT_1(x)}{dx}\bigg|_{x\to 0} = 0 \qquad (3.22)$$

$$(iii) \qquad \kappa \frac{dT_1(x)}{dx}\Big|_{x \to w} = \kappa' \frac{dT_2(x)}{dx}\Big|_{x \to w} \qquad (3.23)$$

$$(iv) \qquad T_1(x)|_{x \to w} = T_2(x)|_{x \to w} \qquad (3.24)$$

First and the second boundary conditions suggest that when we are sufficiently far away from the laser beam, temperature of the crystal equilibriates with the ambient and the temperature under the laser spot is finite. Third and the fourth boundary conditions imply that the temperature at the interface of the supported and the suspended parts are at the same value with incoming and outgoing energies are the same. It is also possible to solve Eq. 3.17 without including the Gaussian heating term and adding the laser power in (3.22). Also, it is possible to include a Newtonian cooling term for measurement under the ambient, which becomes relevant especially when the sample is suspended over a substrate. Under the boundary conditions we obtain the following integration constants:

$$C_2 = C_3 = 0 \qquad (3.25)$$

$$C_4 = \frac{\alpha w_0}{2} \sqrt{\frac{t\pi}{G\kappa'}} \frac{-erf\left(\frac{w}{w_0}\right)}{e^{-wG/t\kappa'}} \qquad (3.26)$$

$$C_1 = T_0 + C_4 e^{-w\sqrt{G/t\kappa'}} + \frac{\alpha\sqrt{\pi}w_0}{2\kappa}\left(\frac{w_0}{\sqrt{\pi}}e^{-\frac{w^2}{w_0^2}} + erf(w/w_0)w\right) \qquad (3.27)$$

Similar to the 2D case, using the thermal solution it is possible to find the resistivity distribution $\rho(x)$. In this case, there is no need to use FEM simulation as the total resistance of the device can be found easily by integrating the resisitvity along the length of the 1D crystal.

Here, I would like to study a hypothetical 1D metallic crystal that is 10 μm long, 0.5 μm wide and 20 nm thick. I assume that the length of the crystal is suspended across the metal contacts, thus the ends are both thermally and electrically in equilibrium with the metal contacts. I assumed heat convection to the ambient and I performed a parametric sweep on the thermal conductivity values to get the temperature distribution on the sample and the resistance change of the crystal. Figure 3.10 shows the plots for each thermal conductivity value running from 1 to 50 W/m K.

3.7 Measuring Thermal Conductivity of Semiconductors

Although intrinsic semiconductors show bolometric response that is well defined with the Arrhenius relation, due to lack of a local heating mechanism other than optical, the use of the bolometry based thermal conductivity measurement method is not straightforward. In this section, I would like to discuss the applicability of the method on 2D semiconductors.

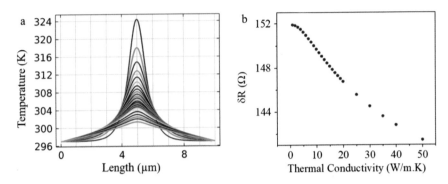

Fig. 3.10 **a** Temperature distribution along the width of the crystal for various thermal conductivity values ranging from 1 to 50 W/m K. As the thermal conductivity increases, the peak temperature decreases but the overall temperature of the crystal rises. **b** Change in the resistance shows unique values for each thermal conductivity value ensuring that the measurement will return a single value

There are various mechanisms for photo-generated currents in semiconductors. For the timescales shorter than a millisecond that is relevant to the timescale of our measurement method, photoconductivity in a semiconductor under bias can result from the formation of non-equilibrium carriers due to light absorption, separation of non-equilibrium carriers due to built-in electric fields or photothermal effects. Each of these effects has to be distinguished from each other to use the bolometric response as a local temperature probe. Moreover, when a 2D material is suspended there might be strain induced built in electric fields that may result in complicated non-equilibrium carrier separation patterns. Figure 3.11 shows the data collected on a multilayer (12 nm thick) WS_2 crystal. In particular variation in the photoconductance at different biases indicates coexistence of various photoconductance mechanisms. Still the calculations for the thermal conductivity gives 8 W/m K which is slightly lower that what has been reported in the literature [6].

We also performed SPCM on MoS_2 crystals. However, we obtained a very interesting zero bias photoresponse over the suspended part of the crystal as shown in Fig. 3.12c. We hypothesize that the unique photoresponse emerges due to the stress induced electric field with in the suspended part of the crystal that results in separation of the non-equilibrium carriers induced by the optical excitation. When the bias is applied, photoresponse is as complicated as the 0 mV bias and extracting the sole bolometric effect is not possible. These initial measurement on the semiconducting materials clearly show the need for very careful measurements to extract the bolometric effect out of other photoconductance mechanisms. However, this may not be a straightforward task and one might be better off with other measurement methods on semiconductors. I would like to emphasize one distinction here regarding the nature of the energy gap. For materials that can be described by single-electron models within the Fermi-liquid theory, considerations I made above are valid. However, for materials such as VO_2 where the insulating gap opens due to strong electronic cor-

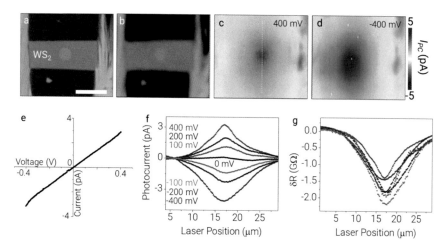

Fig. 3.11 **a** Optical microscope image of a multilayer WS$_2$ crystal contacted with indium needles. Scale bar is 10 µm. **b** Reflection map and **c, d** corresponding photocurrent maps taken at 400 mV and −400 mV respectively. **e** IV plot is mostly linear indicating small barriers at the indium contacts. **d** Line trace taken along the width of the crystal shows the photoresponse through the suspended part of the crystal under different biases. **f** Photoconductance shows slight variation under different biases hinting existence of another mechanism that might be producing current beyond the bolometric effect. Reprinted with permission from Cakiroglu et al. [1]. Copyright (2020) IOP Publishing

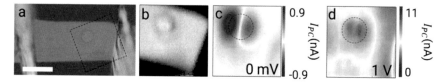

Fig. 3.12 **a** Optical microscope micrograph of a 10 nm thick MoS$_2$ crystal contacted with indium needles. **b** SPCM reflection map and **c, d** corresponding photocurrent maps at 0 and 1 V biases. Notice bipolar photoseponse from the suspended part of the crystal. Reprinted with permission from Cakiroglu et al. [1]. Copyright (2020) IOP Publishing

relations since the thermalization happens very rapidly within less than the length of a unit cell, bolometry based thermal conductivity measurement method will still be applicable in these materials.

3.8 Concluding Remarks

In this brief I tried to introduce the recent literature on the thermal conductivity of 2D materials including our recent contribution. 2D materials field is still an emerging field and for the reasons mentioned in the first chapter, the number of studies in the field is growing exponentially with many papers appearing every day. It is nearly impossible to capture all the literature in a brief of this nature. Thus, I had to narrow down the topic to the single crystalline 2D materials and skipped the literature entirely on multi-crystalline 2D materials and composites of 2D materials. Furthermore, I omitted the discussions that might be relevant to the thermoelectric energy harvesting. Bi_2Te_3 as an instance layered material that shows exceptional figure of merit for the thermoelectric energy harvesting, however I omitted any discussion on such materials out of the scope of this brief. I tried to provide as many details as possible that is relevant to the theoretical discussion of the thermal conductivity in 2D materials and the experimental methods. Moreover, I believe the method we introduced is superior to other available methods in the literature in certain aspects. Thus I tried to provide an in depth introduction of the method with all the details that can not be provided in a short journal paper here. I hope this will help other researchers to understand, apply and improve the bolometry based thermal conductivity measurement method. To conclude, there are still a lot that can be learned on the thermal properties of the 2D materials. For instance, there is no comprehensive study on the thermal properties of the Moire superlattices. Any theoretical and experimental work on such systems may open new opportunities in tunable thermoelectric systems.

References

1. Cakiroglu O, Mehmood N, Çiçek MM, Aikebaier A, Rasouli HR, Durgun E, Kasirga ST (2010) 2D Materials (Dec 2016), 11. https://doi.org/10.1088/2053-1583/ab8048. https://iopscience. iop.org/article/10.1088/2053-1583/ab8048
2. Mehmood N, Rasouli HR, Çakıroğlu O, Kasırga TS (2018) Phys Rev B 97(19):195412. https://doi.org/10.1103/PhysRevB.97.195412
3. Castellanos-Gomez A, Buscema M, Molenaar R, Singh V, Janssen L, Van Der Zant HS, Steele GA (2014) 2D Materials 1(1). https://doi.org/10.1088/2053-1583/1/1/011002
4. Kasırga TS, Sun D, Park JH, Coy JM, Fei Z, Xu X, Cobden DH (2012) Nat Nanotechnol 7(11):723. https://doi.org/10.1038/nnano.2012.176. http://www.nature.com/articles/nnano.2012.176
5. Luo Z, Maassen J, Deng Y, Du Y, Garrelts RP, Lundstrom MS, Ye PD, Xu X (2015) Nat Commun 6(1):8572. https://doi.org/10.1038/ncomms9572. http://www.nature.com/articles/ncomms9572
6. Zobeiri H, Wang R, Zhang Q, Zhu G, Wang X (2019) Acta Materialia 175:222. https://doi.org/10.1016/j.actamat.2019.06.011. https://linkinghub.elsevier.com/retrieve/pii/S1359645419303702

Printed in the United States
By Bookmasters